Education in the Age of Biocapitalism

New Frontiers in Education, Culture, and Politics

Edited by Kenneth J. Saltman

New Frontiers focuses on both topical educational issues and highly original works of educational policy and theory that are critical, publicly engaged, and interdisciplinary, drawing on contemporary philosophy and social theory. The books in the series aim to push the bounds of academic and public educational discourse while remaining largely accessible to an educated reading public. *New Frontiers* aims to contribute to thinking beyond the increasingly unified view of public education for narrow economic ends (economic mobility for the individual and global economic competition for the society) and in terms of efficacious delivery of education as akin to a consumable commodity. Books in the series provide both innovative and original criticism and offer visions for imagining educational theory, policy, and practice for radically different, egalitarian, and just social transformation.

Published by Palgrave Macmillan:

Education in the Age of Biocapitalism: Optimizing Educational Life for a Flat World
By Clayton Pierce

Education in the Age of Biocapitalism

Optimizing Educational Life for a Flat World

by
Clayton Pierce

KH

First published in 2013 by
PALGRAVE MACMILLAN®
in the United States—a division of St. Martin's Press LLC,
175 Fifth Avenue, New York, NY 10010.

Where this book is distributed in the UK, Europe and the rest of the world, this is by Palgrave Macmillan, a division of Macmillan Publishers Limited, registered in England, company number 785998, of Houndmills, Basingstoke, Hampshire RG21 6XS.

Palgrave Macmillan is the global academic imprint of the above companies and has companies and representatives throughout the world.

Palgrave® and Macmillan® are registered trademarks in the United States, the United Kingdom, Europe and other countries.

ISBN: 978–1–137–02782–5 (paperback)
ISBN: 978–1–137–02781–8 (hardcover)

Library of Congress Cataloging-in-Publication Data

Pierce, Clayton.
Education in the age of biocapitalism : optimizing educational life for a flat world / Clayton Pierce.
p. cm.
ISBN 978–1–137–02781–8 (hardback)—
ISBN 978–1–137–02782–5 (paperback)—
ISBN 978–1–137–02784–9 (ebook)
1. Academic-industrial collaboration—United States. 2. Biotechnology industries—United States. 3. Education, Higher—Aims and objectives—United States. 4. Neoliberalism—United States. I. Title.

LC1085.2.P56 2013
378.1'035—dc23 2012028025

A catalogue record of the book is available from the British Library.

Design by Newgen Imaging Systems (P) Ltd., Chennai, India.

First edition: January 2013

10 9 8 7 6 5 4 3 2 1

10/20/14

Contents

Acknowledgments

In many ways, the culmination of a book is more of a historical account of one's relations with a whole host of people, places, and things than an individual accomplishment. Taking this insight as a point of departure, I would like to briefly take time to acknowledge the history of relations that have played a particularly instrumental role in the construction of the ideas, thoughts, and perspectives I have tried to articulate in this book. The first person I would like to thank is Michael Forman for his continued friendship and mentorship—you exemplify the fact that good teachers do more than deliver information but instead awaken the critical agency of students. Douglas Kellner, whose mentorship and friendship has also been invaluable, I would like to express a deep sense of gratitude for your sustained modeling of the critical theory tradition for the next generation. The opportunities Doug has shared with me have been truly generous and educative—never did I dream that I would study and write about Herbert Marcuse with the most renowned Marcusean scholar in the world. Sandra Harding has also been particularly influential in the development of my research and approach to understanding the ways science and technology harbor cultural lessons that teach us more than what can be explained through traditional understandings of objectivity. Her work remains an inspiration and guide. I would also like to make special mention of Eugene Victor Wolfenstein who introduced me to the thought of W. E. B. Du Bois (as well as guiding me through the philosophical labyrinth of Hegel's *Phenomenology of Spirit*) and was the most artful pedagogue I have had the privilege of learning from. Your incredible presence in this world is surely missed by many.

This book would also not have been possible without the wonderfully brilliant graduate students of the Education, Culture, & Society Department at the University of Utah whom I have had the pleasure of working with and learning from over the past five years. In particular, I thank the students who have been part of my Neoliberalism and Schooling, Ecojustice Education in the Age of Biocapitalism, and W. E. B. Du Bois seminars for helping fertilize many of the ideas and concepts that I have been able to develop in this book. Your collective knowledge and effort demonstrates that multitudinal labor is far more powerful than individual effort. I want to also acknowledge my colleague

Leticia Alvarez-Gutiérrez who generously invited me to collaborate with her on developing community-driven projects with students, teachers, and families. I would also like to thank my colleagues Frank Margonis, Harvey Kantor, Edmund Fong, Richard Kahn, and Tyson Lewis for reading and commenting on early drafts of chapters for this book and for partaking in conversations of and about many of the ideas that needed gestation. My deepest gratitude must also be extended to the excellent group of graduate students Christy Call, EnginAtasay, Greg Bourassa, Graham Slater, and Erik Bowen who read and commented on various parts or the whole manuscript. I also am very appreciative of Kenneth Saltman who has been extremely supportive throughout the whole process of bringing out this book, from prospectus to final version of the manuscript. I am very pleased this book will be the first in Ken's series with Palgrave Macmillan that will surely be an original and groundbreaking series.

Finally, I want to acknowledge my heartfelt and sincere appreciation for my parents Terry and Lynn who have been a continual source of love and support throughout my life. This book would not have been possible without them and especially the sense of wonder and respect to the natural/nonhuman world they instilled in me at an early age. Also to my sister Jennifer and niece and nephew (Kealia and Kohala) whose love is a source of great joy in my life. Lastly, my partner Dolores Calderón Estrada and our beautiful daughter Vivianna Calderón Pierce bring me grounding and love each and every day, which is truly a gift. Both remind me that the important things in life start with loving relationships that teach us to continually learn to be better human beings. Lola and Vivianna make it easy for me not to forget this source of wisdom.

INTRODUCTION

Biopolitics and Education: A Return to the Question of Life and School

> America's future economic growth and international competitiveness depend on our capacity to innovate. We can create the jobs and industries of the future by doing what America does best—investing in the creativity and imagination of the people. To win the future, we must out-innovate, out-educate, and out-build the rest of the world.
>
> Our innovation strategy begins with critical foundations: education, scientific research, and infrastructure. First, we must create an educational system that is internationally competitive and innovative in preparing our workforce for our increasingly knowledge-intensive economy. Second, we must invest in scientific research to restore America's leadership in creating the scientific and technological breakthroughs that underpin private sector innovations. Finally, we must invest in a first-class infrastructure that moves people and ideas at twenty-first century speeds. These are the building blocks of an innovation strategy that will lead America to a more prosperous future.
>
> —A Strategy for American Innovation: Securing Our Economic Growth and Prosperity

National Economic Council (2011)

Over the past few years a startling number of films have taken schooling in the United States as its cinematic subject. *The Lottery, The Cartel,* and *Race to Nowhere* are just three of the most notable documentaries that explore educational issues ranging from school choice, declining student performance in math and science, to the high-stakes culture of test-driven curricula that continue to shape life in public schools across the nation. However, it is perhaps Davis Guggenheim's (director of Al Gore's *An Inconvenient Truth*) *Waiting for Superman* (2010) that has garnered the most attention. Undoubtedly, much of the buzz around Guggenheim's film had a lot to do with its release date which coincided with the start of a new school year in a country still mired in economic crisis. Yet why did Guggenheim's film resonate with so many, not only with well-meaning democrats but also with conservatives who have long been waiting for the body of public education to wither away and die? Part of the answer to this question resides in the film's overriding thesis that teacher performance and specifically the ability to remove ineffective teachers from the classroom is one of the biggest barriers to building a nation of high-achieving schools capable of producing students who are prepared to compete in a high-stakes, knowledge-driven global economy.

Beyond the fact that *Waiting for Superman* functions as a form of media spectacle for the Obama administration's education reform strategy Race to the Top, one that targets teachers' unions and tenure, the underlying message being sold is unmistakable: schools and, in particular, student achievement are inextricably linked to the country's economic stagnation and national security concerns. In this sense, the arc of Guggenheim's film signals a now familiar sky-is-falling narrative about the public schooling system in the Unites States. Good test scores in subjects such as math and science, in other words, are not simply an academic matter of concern. Rather, what is becoming clearer in what I call the growing neo-*Sputnik* narrative driving education reform in the United States today is how deeply implicated the educational population of the United States has become with strategies of economic recovery and the maintenance of imperial dominance. Sadly, the register of educational health pointed to in Guggenheim's film, reflective of the neo-*Sputnik* fervor shaping educational discourse and policy reform in the United States today, is that of economic vitality and, specifically, the nation's ability to reenergize its high-tech human capital base through privatized charter school ventures, corporate/public partnerships, and the application of stricter disciplinary practices to educational spaces and governance. The reinvention of the US public school in Guggenheim's vision rests largely on reform models conceived by Bill Gates, Merck Pharmaceutical, and a multitude of other commercial and corporate educational enterprises not interested in expanding the capacities of individuals and communities to engage with the host of economic, ecological, and social crises now shaping both the present and future. Instead reform and policy driven by stakeholders who are invested in the future of biocapitalism are more concerned with the expansion of the market through new discoveries involving the commodification and exchangeability of forms of life.

In another recent film examining contemporary life in schools, an entirely different interpretation from Guggenheim's is offered. Cevin Soling's *War on Kids* (2009) depicts public schools in urban and suburban contexts as more akin to the prison industrial complex than a site where bad teachers protected by powerful teachers' unions continue to be an obstacle to making US schools more competitive in what Thomas Friedman has called the "flat world". Here, schools in both neglected and underserved inner cities and middle class white suburbs are presented as environments of high surveillance and punitive measures, a context where students' daily lives are impacted more by a panoply of metal detectors, cameras, security guards, behavior-altering pharmaceuticals, alarm systems, and chain-link enclosed computer labs than by a teacher who is not doing her or his job effectively. What Soling's film provides is a more accurate ethnographic account of life in schools in the United States, where, as Michel Foucault has suggested in his work, technologies of management and control are actively intervening into the field of educational life. Schools, put differently, have become an important part of the biopolitical landscape associated with neoliberal strategies of governance for optimizing the population in a flat world economic arrangement. Where these two films converge, albeit from different perspectives, is the subject this book sets out to investigate: the manner in which life in schools has become more deeply integrated with perpetual economic crisis and the promissory futures created through the increasingly biocapitalist arrangement of key economic industries.

To begin mapping out an educational terrain where *life* is more and more becoming the target of powerful regimes of control and production, this chapter introduces a biopolitical framework that I use throughout the book to examine life in schools in the biocapitalist era. As I argue in the following sections of this chapter, schooling and education more generally cannot be understood outside of the economic renewal projects and national security crises that are framing educational reform debates in the United States today. Given the centrality of educational life to such powerful projects, a biopolitical analysis is a particularly well-suited tool for situating and understanding what has become the political object of both the economic and security projects underlying neo-*Sputnik* educational reform driven by biocapitalist imperatives: the productive potential of life. The goal of this chapter is thus twofold. First, it provides an introduction to biopolitics as an interpretive and critical framework for understanding the dominant economic and security strategies and practices that I suggest are intervening into more and more aspects of what constitutes the production of educational life; or, how subjectivities are formed in neoliberal spaces of learning. Second, this chapter also lays the groundwork for examining what I am calling *extractive schooling*, or the process by which educational vitality has become a mineable good within the neo-*Sputnik* terrain of educational discourse and reform strategies. I am using the term extractive here in two senses: educational life, as it relates to neo-*Sputnik* discourses and practices, can best be understood as (1) a field of latent value that (2) is actively being mined by a host of apparatuses taking their cue from the promissory future that biocapitalist industries and stakeholders imagine and seek to create.

One of the ultimate claims this book makes is that in order to reach a better understanding of the deeper projects energizing education reform and the general neoliberal drift that public schooling has been subsumed within the United States over the past 30 years, it is perhaps most instructive to look at the ways in which life (in both the biological and the sociological sense of the word) is being produced and *valued* at both the discursive and material levels of educational debates and practices emanating from the general milieu of economic crisis and security-oriented policies now driving educational reform strategies. In other words, the national project to "out educate" the rest of the world in a "race to the top" exemplifies a particular typology of educational life that I excavate and examine throughout the chapters of this book. It is a type of educational life, I argue, that has deeply entangled the education of subjects with both crisis management techniques of capital and neoliberal conceptions of the individual as an investable and manageable body. Within this new educational anatomy, shaped by the needs of a quickly growing biocapitalist economy, the dissolution of restrictive barriers to the commodification of more aspects of life is of greater concern than working toward alternative educational futures based on socially just and ecologically healthy communities.

One way to ascertain some of the root features animating responses to the current neoliberal educational reform direction of this country is to start thinking like genetic scientists. That is, much in the same way the genetic sciences have figured out how to manipulate and manage life at the cellular level, educational theorists, educators, and communities also need to recognize that greater areas of biological *and* social life are being enlisted into new productive regimes that have emerged from the highly speculative and volatile caldron of bioindustrial growth. As I argue throughout this book, the productive needs of the lab and the bioeconomy in general require particular forms of educational life in order to function and expand. Yet it is also from within this emerging terrain that practices of resistance and alternative forms of educational life must be redrawn if life is going to be redirected from projects oriented toward the further economization of human and nonhuman forms of life to healthier and just ones. One of the overall goals of this book, therefore, is an attempt to help make these biopolitical lines clearer and to point to an alternative direction in which new educational life can sprout. To begin to lay out some of the ways an alternative biopolitics of education can be generated in different pedagogical contexts, we first need to look at the origins and key attributes of biopolitical critique in order to see what it offers over other interpretive models.

From Biopower to Biocapital: Foucault and the Politics of Life

The term biopolitics originated in the research Michel Foucault conducted on the ways sovereign power underwent a variety of qualitative changes throughout the modern period. Since the publication of Foucault's Collège de France lectures in English over the past ten years, biopolitical perspectives have quickly spread in the areas of social and political theory, philosophy, and science and technology studies. Two of most influential and productive areas that have helped develop the framework of biopolitical analysis that I draw upon in this book to analyze the nexus of biocapitalism and education are the work of political theorists Michael Hardt and Antonio Negri, and science and technology theorists who have done the most in charting the rapidly expanding landscape of biocapital. These two genealogical branches of the biopolitical tradition offer a framework I adopt in this book to understand how (1) schools can be thought of as productive sites of subjectivity that are (2) being shaped by the needs of an economic paradigm that banks its future on scientific and technological advances designed to capitalize on forms of life and (3) to think through what an alternative biopolitics of educational life might look like in such a context. In order to begin to put these pieces into motion through my examination of extractive models of schooling, it is necessary to first clarify the analytics constituting the framework of what has become known as biopolitics. The best place to start is with the "birth of biopolitics" in the work of Michel Foucault.

In his 1975–1976 lectures titled Society Must Be Defended (2003), Foucault offered his first sustained account of biopolitics that built on earlier sketches of the concept he introduced in the *History of Sexuality Volume I: An Introduction* (1978). Through his study of modern forms of disciplinary power over bodies subjected to institutional knowledge systems and practices associated with places such as the prison, hospital, factory, military, and school, Foucault stretched his analytic of biopower beyond the scope of the individual to encompass a model of state power organized and directed toward the control and regulation of populations. Pointing to the Nazi state as an absolute expression of biopolitics in the modern era, Foucault (2003) postulated that "in Nazi society we have something that is really quite extraordinary: this is a society which has generalized biopower in an absolute sense, but which has also generalized the sovereign right to kill" (260). Embedded within the dual projects of state racism and state capitalism, the entire population within Nazi society, according to Foucault, fell under a type of biopower that enlisted individuals into the Nazi industrial war machine or removed them from the population through death or encampment. Here, biopower, in Foucault's articulation, had achieved its apex in the modern era: life in general was subject to a historically unique practice and theory of politics that took human vitality and its regulation as its object.

In his analysis of the emergence of a biopolitics in the modern state and its most potent example in Nazi society, Foucault lays out for the first time his "diagram of power over life" where "one pole of biopower focuses on an anatomo-politics of the human body, seeking to maximize its forces and integrate it into efficient systems" and where the other "pole is one of regulatory controls, the body imbued with the mechanisms of life: birth, morbidity, morality, longevity" (Rabinow and Rose 2006, 196). In other words, for Foucault, the Nazi state combined absolute disciplinary power through its brutally efficient bureaucracy, the SS, and media propaganda, with the absolute biological regulation of the population (control over procreation, heredity laws, compulsory sterilization, and social programs to exterminate "impure" races). Yet Foucault did not see the Nazi example as a historical anomaly; in fact, he argued that all modern capitalist and socialist states exhibit the very same tendencies to control and regulate their populations, though perhaps not in such extreme terms.[1] What biopolitics therefore generally means in Foucault's (2003) work is a type of politics that "deals . . . with the population as political problem, as a problem that is at once scientific and political, as a biological problem, as power's problem" (245).

It is important to make clear three main attributes of biopolitics enumerated by Foucault in his 1975–1976 lectures as distinguishable from earlier forms of sovereign power embodied in disciplinary institutions and the rationalities produced therein. For Foucault, disciplinary institutions and their attendant "regimes of truth" create a type of biopower geared toward the internal ordering of individuals by enlisting individuals into practices and habits that work through bodily techniques of control imbued with moral and ethical pedagogies of the self. Here the confession, panoptic surveillance within the prison, or the schoolmasters' dictations in the classroom or lecture hall function as key disciplining actors that influence and teach an ethic of self-care to guide the conduct of individuals in society and at home. Different from the disciplinary expressions of biopower that dominated the early modern period, which were largely excercised through juridical means, biopolitics for Foucault is a shifting of focus of sovereign power from the individual to the population. Such a transition in sovereign power's nature, according to Foucault, began with the state's recognition of life within a territory or space as a political problem, something to be managed and controlled through a whole new series of statistical and measurement techniques and practices of governing. Here Foucault (2003) gives historical examples of the state from the eighteenth century onward, directing its political interests toward matters of morbidity, birth, city sanitation, disease prevention, and generally the question of life in densifying urban settings. The first aspect of biopolitics Foucault delineates is thus a qualitative transformation in both the scope and focus of biopower.[2]

The second feature of Foucault's biopolitics hinges on the issue of scale: more precisely, how random events and trends in the population of a state or territory are integrated into the processes of knowledge production (or discursive regimes) designed to intervene and order life within a population. The census, for example, represents what Foucault means here by a study of "phenomenon over time" where agencies and governments count and sort the population of a state using different demographic rationales and statistical measuring techniques to form a calculus of governance. From studies of the population such as the census, data are used to adjust and create new technologies of control and management such as the establishment of voting districts or the institution of "sundown" laws to make the night white in many US cities and towns for the better part of the twentieth century in the United States. Or, in the context of schooling in the United States, studies and techniques used for redistricting schools would also be an example of managing and controlling segments of the population through the structuring of access to schools via district boundaries. This type of biopolitical regulation is especially evident in cases where redistricting either unites or separates segments of the educational population that have historically fallen along racial and class lines in US cities as well as rural area (Boger 2002).

The third and connected component of biopolitics is what Foucault sees as the development of "security measures" or what Hardt and Negri (2000) have termed "technologies of control" in their interpretation of Foucault's understanding of biopolitics. This feature of biopolitics can perhaps best be understood through its practical or pragmatic quality. For Foucault (2003), "it is therefore not a matter of taking the individual at the level of individuality but, on the contrary, of using overall mechanisms and acting in such a way as to achieve overall states of equilibration or regularity; it is, in a word, a matter of taking control of life and the biological processes of man-as-species and of ensuring that they are not disciplined but regularized" (246). An example of such a technology of control associated with the evolution of biopolitics in modern society that fits with Foucault's framework are the compulsory sterilization programs that were a part of the eugenics movement in Europe and the United States in the first part of the twentieth century. As Angela Davis (1983) has pointed out, the targets of sterilization laws and programs in the United States were largely American Indian and African–American women as well as prison populations throughout the country that are disproportionately filled with people of color. Through racial hygiene laws such as compulsory sterilization, states across the United States were involved in the biopolitical regulation of portions of the population that were deemed problematic, criminal, or impure by the government that called for a violent intervention to "take control of life and the biological processes of man-as-species."

Taken together, biopolitics in Foucault's thought at it most general level can, therefore, be understood as a tracing of the movement of sovereign power over life from the one, a king or a ruler of a people, to a diffuse set of power relations that target and regulate the life of an entire population within the boundaries of a state or territory. At this point, it would perhaps be beneficial to look to a historical example rooted in a well-known educational debate of the twentieth century as a way to draw out Foucault's notion of biopolitics a bit further. Particularly how biopolitics function through "rationalities" or "programs" that emerge from the state's problematization of phenomena in the population around issues such as health, policing, education, or incarceration. As I argue below, the school has been one of the most effective and destructive technologies of control within the history of the United States. American Indian boarding schools and segregated schooling, for instance, were not idiosyncratic to the historical landscape of education in the United States; rather, these sorting and regulatory institutions provided a desired function for the state, what W. E. B. Du Bois called a system of caste education.

Du Bois's well-known debate with Booker T. Washington and the "Tuskegee Machine" captures many features of what Foucault termed biopolitics. Indeed, Du Bois recognized that the stakes of the debate around vocational versus liberal arts education rested on the question of how schools and educational institutions generally would play a central role in the regulation of the African–American population within US society. That is, if schools and institutions of higher education in the pre-Civil War period regulated the African–American slave and freed populations through legal exclusion, forced plantation encampment, or death, Du Bois (1930) also viewed the rise of the vocational model of schooling in the Reconstruction period along the same continuum. For Du Bois, in other words, schools sought to assimilate African–Americans "into the great maelstrom of the white civilization surrounding us" where "we have been inevitably made part of the vast modern organization of life where social and political control rests in the hands of those few white folk who control wealth, determine credit and divide income" (191).[3]

Wrapped up in the question charging the debate between Washington's accommodationist (vocational schooling) and Du Bois's emancipatory views (broad liberal arts and practical arts), therefore, is a tension that has at least three important biopolitical features I want to accentuate here: the problematization of a segment of the population through racial science; the establishment of technologies of control by the state to promote and enforce habits of obedience and docility in the African–American population beneficial to industrial labor; and the active maintenance of whiteness as the normalized measure of citizenship in US society. What also becomes clear in reading Du Bois through a biopolitical lens is that Foucault was not the first to recognize how institutions such as the school and knowledge systems such as eugenic science shaped, managed, and intervened into the life of human beings. Du Bois also recognized that the stakes in his struggle against the Tuskegee Machine was one of life and death, both in the physical and social sense of the term. In fact, what Du Bois called the "caste education system" operating in the United States can be seen as a biopolitical critique of schooling before the term appeared in Foucault's work in the mid-1970s.

As noted above, the first element of biopolitics according to Foucault involves the problematization of the population; that is, the state's ability to make the population a political problem to be studied, measured, managed, and controlled in society through a variety of technologies of control. Underlying the Du Bois–Washington debate was indeed the question of problematizing African–Americans as a population group, specifically how African–Americans should be educated. During the first part of the twentieth century in the United States, African–Americans, as Du Bois himself notes in *Dusk of Dawn* (1995) and elsewhere, were considered by the white ruling caste to be a distinct social problem, a "disease" that had to be managed and controlled so as not to infect the health and purity of the white population. Du Bois (2007, 1995, 2009), in fact, discussed throughout his work, perhaps most famously in *The Souls of Black Folk*, the existential condition of being considered a "problem" in his own society, an individual who was considered as neither a US citizen nor a part of the imagined exceptionalist history maintained in white supremacists accounts of the founding of the United States. Du Bois's (1995, 1996) own early sociological study of the African–American section of Philadelphia, for example, reflects this tension through the fact that his research was funded and directed by the University of Pennsylvania and also supported by the white city leaders who were desperately looking for a way to deal with what they perceived to be the growing problems of poverty, crime, alcoholism, and prostitution in the Seventh Ward of Philadelphia.

In many ways, Du Bois's seminal work *The Philadelphia Negro* can be seen as a direct confrontation with eugenic racial views that had become the leading scientific explanation used in social policies to demark and find solutions to problems of "racial hygiene" during the first part of the twentieth century in the United States and Europe as well as their colonial holdings. As David Levering Lewis (1993) remarks, "Du Bois knew his sponsors held a theory about the race he studied. The city was 'going to the dogs because of the crime and venality of its Negro citizens.' 'Something is wrong with a race that is responsible for so much crime,' the theory ran, and 'strong remedies are called for' "(188–189). Backed by eugenic beliefs, the funders of the study Du Bois undertook hoped to collect data and research that could help establish systems of control for better managing what they viewed as an inferior and potentially dangerous social problem represented in the African–American population of Philadelphia.[4] Du Bois, for his part, was still at a point in his career where he believed that the truth and clarity provided by scientific research could erode the endemic racial views held by whites that kept in place the problem of the color line in the United States. As Du Bois (1995) himself recalled, "Philadelphia wanted to prove this by the figures and I was the man to do it. Of this theory back of the plan, I neither knew nor cared. I saw only a chance to study an historical group of black folk and to show exactly what their place was in the community" (58). Du Bois's study of the Seventh Ward of Philadelphia along with his battle with the Tuskegee Machine, despite his best efforts, precisely reflects important attributes of biopolitics. The first is present in how the city of Philadelphia actively constructed the ontological condition of being African–American as tantamount to a problem, something to be studied, measured, classified, and controlled through rationalities (a study of the city's African–American population) that also normalized racial scientific views within society and in particular, how different segments of the population are to be governed.

The second biopolitical feature I am suggesting undergirds the study undertaken in *The Philadelphia Negro* and the Du Bois–Washington debate in general is the recognition by state and federal governments that freed slaves required institutional mechanisms or governmental interventions (such as the failed Freedmen's Bureau and free public schooling) for facilitating social survival after the fall of legalized slavery in the United States. Here it can be argued that state, city, and county governments exhibited what Foucault indentifies as a signature attribute of biopolitics: a propensity to identify and deal with a particular phenomenon emerging from within the population. In this case, the phenomenon to be managed precipitated from the dissolution of the institution of slavery which initiated the need to develop new methods in which to preserve white supremacy as the primary model of citizenship and the privileges of property ownership and enfranchisement it protected as a "legal" form of social control (Du Bois 1999, Olson 2004).[5] The series of social and economic reforms during the Reconstruction period, viewed from the perspective of state interventions into the social life of African–Americans, it could thus be argued, signified a governmental attempt to manage a wide-scale biopolitical problem that centered on this problematic: how to regulate life in a post-slavery society while maintaining white supremacy as the governing archetype of political subjectivity. Given the fact that these transitional institutions were underfunded and left to the control of the white ruling elite (intentially filled by ex-Confederate politicians), something Du Bois pointed out with regularity in his many critiques of the African–American educational system in the United States, the problem of African–American social inequality was dealt with by federal and state governments after the Civil War represents a prime example of how biopolitical goverining strategies deal with an emerging phenomenon in the social organism that threatens dominant power relations such as white supremacy and industrial capitalism. Regulating the life of the African–American population in the United States as a way to manage the crises of white supremacy and the slave economy brought on by emancipation consisted of strategies such as the caste education system that ensured hierarchical and unequal relations could be maintained along racial and class lines in state run schools.

Finally, a third biopolitical aspect present in the example of the Philadelphia study as well as the Du Bois–Washington debate is the establishment of types of technologies of control over the African–American population through state and federal governments. In this sense, what Du Bois called the Tuskegee Machine or empirical investigations of racial groups in a city funded by white interests can also be understood as a mechanism by which white society "tak[es] control of life and the biological processes of man-as-species" and "ensur[es] that they are not disciplined, but regularized" (Foucault 2003, 247). Thus in looking at how African–American education was being constructed as a means to uphold white supremacy in the United States through what William Watkins (2001) has called the "white architects of black education," the three biopolitical pieces of the Du Bois–Washington debate and the study of the African–American population in Philadelphia come together: (1) the establishment of the emanciptated African–American population as a political problem by the state; (2) the production of knowledge regimes and institutions (the study of urban context with eugenic science beliefs, the Freedmen's Bureau, official historical accounts valorizing white supremacist values in school curriculum) that legitimatize forms of intervention and regulation of the African–American population; and (3) the creation of technologies of control (segregation laws, indentured farm work, forms of social terror such as lynching, miscegenation laws, segregated schooling system, etc.) to maintain white supremacy as the normalized model of civic life in US society. Moreover, as Du Bois brilliantly lays out in his seminal *Black Reconstruction* (1999), one of the greatest beneficiaries of state biopower that utilized race to fragment and divide society such as caste education was industrial capital. That is, one of the primary ways the growing need for a large industrial labor pool in the late nineteenth and early twentieth century in the United States was achieved was through the pitting of white workers against black, ultimately bringing down wages and increasing profit for industrialists, who, not coincidently, were the primary funders and directors of the Tuskegee Institute and the vocational system of schooling established for African–Americans during this period (Du Bois 2011, Roediger 1991, Olson 2004).

What we can ultimately say then about Du Bois's biopolitical critique of education is that it centers on the maintenance and perpetuation of a caste education system in the United States. In reassessing the conditions at Fisk University and the overall status of black colleges in the United States in the 1930s, Du Bois clearly understood the status of African–American education along biopolitical lines. Du Bois (2001) remarks, "Biologically we are mingled of all conceivable elements, but race is psychological, not biology; and psychologically we are a unified race with one history, one red memory, and one revolt. It is not ours to argue whether we will be segregated or whether we ought to be a caste. We are segregated; we are a caste. This is our given and at present unalterable fact. Our problem is: How far and in what way can we consciously and scientifically guide our future so as to insure our physical survival, our spiritual freedom and our social growth? Either we do this or die. There is no alternative" (130–131). For Du Bois, the role of Historically Black Colleges and Universities (HBCUs) in the era of Jim Crow is understood as potential sites of biopolitical resistance, in that he interprets their broader aims as promoting emancipatory forms of democratic life in a racially segregated and economically unequal society. Thought of from a biopolitical standpoint, then, the integration or assimilation of African–American communities into industrial or "Northern" models of schooling during the Reconstruction period and beyond, a key area of focus in Du Bois's analysis of caste education, reflects the regulatory type of biopower that schooling performs in white supremacist and capitalist society. In other words, in Du Bois's critique of caste education in the United States, schools and institutions of higher education act as technologies of control in that they intervene into the social life of African–Americans through governmental regulation and control in a way that ensures the existence of what Du Bois called the "white world" and the form of political life it both allows and disallows.

Yet Du Bois's analysis of caste education also highlights another key feature of Foucault's biopolitical framework: the concept of governmentality. As an important bridging concept in Foucault's formulation of biopolitics, governmentality, or what Foucault also called the "art of governing", is a crucial concept because it represents in concrete and historical terms how institutional and state apparatuses cultivate "ethics of self care" in the subject (disciplining biopower) that are connected to techniques of control (regulating biopower) administered through the state. In the context of the Du Bois–Washington debate, governmentality, in its most general sense, can perhaps best be understood through forms of biopower such as the caste education system that functioned to control and manage African–American individuals and populations within the political parameters of a white supremacist society. African Americans, within a caste education system, were subjected to curriculum and teaching practices that promoted and disciplined easily translatable skills and habits to menial and unhealthy labor needs while voting, housing, and labor laws helped regulate a segregated society and thus white supremacy. It is not suprising then that part of Du Bois's argument against Washington and white elite control of African–American educational institutions was that segregated schooling was also a method for internally governing folks of color through white supremacist values and notions of civic life that evolved from the colonial and slave history of the United States. For example, as Joel Olson (2004) and others have pointed out, labor and housing ordinances (governing the ways people work and live together) compelled individuals and groups such as Irish or Italian immigrant communities to embody white ideologies and habits of civic life which, in turn, helped to uphold and maintain white citizenship as the hegemonic form of political life (Roediger 1991, Olson 2004, Ignatiev 2008).

In choosing to profit from the restriction of African American and other communities of color from certain housing or work options as opposed to viewing these forms of oppression as a common problem shared between the overwhelming majority of all groups, what Du Bois called the "psychological wages of whiteness," white immigrant groups benefited from racism directly by participating in the policing, rioting, and violence against people of color—acts that socially demarcated these groups as distinct from darker skinned communities. Du Bois was thus one of the first political theorists to recognize how white supremacist society participates in maintaining political structures and knowledge institutions that operate in part through governing individuals and communities of color, as well as white folk's internal understanding of themselves and others. Du Bois (2001) lucidly lays out how such internal forms of governmentality play out in the subjectivity of diverse groups of individuals:

> Japanese, Chinese, Indians, Negroes; and, of course, the vast majority of white folk; have been so enthused, oppressed, and suppressed by current white civilization that they think and judge everything by its terms. They have no norms that are not set in the nineteenth and twentieth centuries. They can conceive of no future world which is not dominated by present white nations and thoroughly shot through with their ideals, their method of government, their economic organization, their literature and their art; or in other words their throttling of democracy, their exploitation of labor, their industrial imperialism and their color hate. (158–159)

Du Bois's analysis of caste education within white supremacist society and particularly the ways institutions, policies, and laws create an educational field of coercion in which subjectivities develop is an idea that connects strongly with Foucault's concept of governmentality. Similar to Du Bois's understanding of how schools participate in the production of subjectivities invested in whiteness as the measure of political and social standing, Foucault's study of governmentality is also concerned with practices and rationalities that aim to order the "conduct of conduct" of individuals through strategies of intervention employed by the state. Governmentality in Foucault's framework of biopolitics, in other words, is a concept that seeks to understand precisely how subjects are formed within the existing power relations in society through instances of micropower aimed at ordering the structures of social life within which individuals act and make decisions in daily life. Foucault's notion of governmentality is also an important dimension of biopolitics because of its focus on the social mechanisms geared toward the production of individual habits, behaviour, and desires compelled through institutional and sociopolitical techniques of control. Schooling is a particularly apt institutions in which to view how governmentality structures the ontological field where subjectivities of individuals develop.

Governmentality for Biocapitalist Production: Optimizing the Value of the Educational Population

As Thomas Lemke has pointed out in his work charting the development of biopolitics in Foucault's thought, governmentality can be thought of as the "missing link" between Foucault's two primary areas of study: the "genealogy of the state" and the "genealogy of the subject." Lemke (2002) situates the concept of governmentality in Foucault's work in such a way "because between these two research interests is the problem of government. It is a link because Foucault uses it exactly to analyze the connections between what he called technologies of the self and technologies of domination, the constitution of the subject and the formation of the state" (2). Governmentality thus serves as the framework through which biopower operates; it is in short the field of action that compels the individual to act by facilitating an internalization (or subjectification) of rationalities or "regimes of truth" that emanate from legal, health, or educational apparatuses of the state, for instance. In this sense, the study of government concerns " 'the conduct of conduct': that is to say, a form of activity aiming to shape, guide, or affect the conduct of some person or persons" and is interested in uncovering "the relation between self and self, private interpersonal relations involving some sort of control or guidance, relations within social institutions and communities, and, finally, relations concerned with the exercise of political sovereignty" (Gordon 1991, 2).

Foucault's study of historical forms of governmentality, however, must also be read within his genealogical analysis of what he called the "art of governing" from antiquity, through the middle ages, and into the modern period in Europe. Of particular importance to Foucault in this last phase in the charting of governmentality were liberal strategies of statecraft developed in eighteenth-century Europe. For the purposes of this chapter and throughout the chapters of this book, his examination of post-World War II models of neoliberal economic policies designed to dissolve Keynesian economic practices through radical free-market reform strategies is particularly helpful (Gordon 1991, Dean 2010). In his study of early twentieth-century German neoliberal economic theory (Menschenökonomic) and the Chicago School's more radical version developed after World War II in the United States, Foucault's analysis of governmentality points to the state's increasing ability to connect political concepts such as freedom and autonomy to a decidedly economic origin of the human species: *Homo economicus*.[6] In a world carved out for *Homo economicus* health care or education in societies guided by neoliberal models of governmentality are cast as an individuated practice of investment and entrepreneurial acumen. As a result, governmentality practices for Foucault teach an ethics of the self care through the power relations embedded in institutional actors such as the US health-care system's ability to sort and administer treatments and care through vast and many times unnavigable bureaucratic networks, or through student loan structures that coerce one's access to education to become an act of financialization where individuals' educational cost spans the better part of a lifetime. Or consider the credentialing process for teachers that is many times built around strict procedures and directives that administer prepackaged lessons attached to strong performative measurement tools to an educational population. Each of these institutional and policy structure systems also performs, according to Foucault's concept of governmentality, an important pedagogical function in society: they help cultivate over time an "ethic of self care" within the individual by promoting free-market values and entrepreneurial decision making as the natural and normalized approach to health, education, and other common concerns of people in society. One's health, in other words, is not a sociopolitical concern so much as it is a matter to be dealt with by an individual and his or her relation to the insurance company, employment status and benefits available, hospital choice, drug therapy options, or an understanding of patient rights and types of advocacy one can pursue to increase one's chances of better treatment. The transitioning of social risks brought on by inequitable economic and legal structures to the individual, in other words, is a primary goal of neoliberal forms of governmentality that Foucault's analysis rightly stresses.

Yet one of the biggest strengths of Foucault's (2004) analysis of governmentality in neoliberal society is how it "finds itself in a situation in which it has to refer to the economy as a domain of naturalness; it has to manage populations; it also has to organize a legal system of respect for freedoms; and finally it has to provide itself with an instrument of direct, but negative, intervention, which is the police [understood by Foucault as police forces and diplomatic-military apparatuses]" (354). What's more, in the radical free-market society governed by ethics of privatization and economic competition based on artificial scarcity, governmentality within Western states, Foucault suggest, is also increasingly concerned with "the forces and capacities of living individuals, as members of a population, as resources to be fostered, to be used and to be optimized" (Dean 2010, 20). Here, the connection between Foucault's two areas of study on modern forms of sovereign power (discipline and control forms) can be seen operating in unison within his concept of governmentality. Namely, Foucault's analysis of governmentality suggests that the subject living within matrices of neoliberal governmental strategies finds him or herself in a social environment constantly exposed to forms of market-based disciplines that normalize the total economization of life. In other words, life in neoliberal social and cultural settings is subject to forms of biopower calibrated to shape and inform the habits and decisions of individuals by increasingly economizing more aspects of social life *and* how individuals think of themselves within society as entrepreneurial actors.

Here, one of governmentality's primary qualities as a lens in which to view power relations in society is apparent and can be distilled broadly as "a range of forms of action and fields of practice aimed in a complex way at steering individuals and collectives" (Bröckling et al. 2011a, 1). Witness, for instance, the relentless barrage of media and cultural images promoting online colleges, or flexible course schedules offered by a vast number of university and corporate colleges such as DeVry or University of Phoenix. In their effort to increase consumer choice and reinforce the idea that the more education one has the better one's income stream will be, for-profit colleges are paradigmatic neoliberal actors in the educational landscape of the United States. Similarly, consider the manner in which the student loan structure in the United States operates simultaneously as a form of neoliberal financialization (student loans themselves are a highly profitable investment for the banking industry) and as the primary means by which individuals are able to attend and receive degrees from institutions of higher education. Here, freedom and educational opportunity are bracketed within a framework of neoliberal financial apparatuses and a normalized societal value of education as a consumptive good to be purchased on the free market by entrepreneurial individuals seeking to invest more into their lives as a means to increase their value as a human capital commodity. In both of these examples, freedom is translated into a free-market, individual choice, while education itself, as a commodity, commands an ethic of ownership, self-investment, and continual reinvestment to attain the social value it promises to bring.

As such, Foucault's analysis of neoliberal economic theory reveals how, translated into social policy strategies, neoliberal forms of governmentality aim to "optimize the life of some, while disallowing the life of others" (Dean 2010). That is, at the level of population, neoliberal forms of governmentality are very much concerned with regulating the overall health and productive potential of life in a territory or state, unevenly investing in the lives of some while divesting in others. Looked at through the lens of governmentality, for example, the education of the individual is situated in a modifying relationship between the health and well-being of one's social life via educational investments as well as to the health of capitalist growth. This tension existing between capitalist expansion that relies on high yields of human capital on the one hand and the demands on individuals to invest and optimize themselves on the other epitomizes neoliberal governmental strategies of the state. There are at least two important registers in this power relation between education in the age of biocapital and the production of subjectivities that this book is interested in examining. First, education in neoliberal societies is generally understood to be a site of future economic potential, both at the individual and social levels of investment. For example, the human capital model of education in the United States, the subject of chapters 1 and 2 of this book, primarily situates education as an engine of economic growth in a highly competitive global environment. Such a view has been steadily increasing since World War II and evidenced in the trail of policy interventions into education reform from the Sputnik National Defense Education Act of 1958, *Nation at Risk*, No Child Left Behind to the current neo-*Sputnik* policy reform recommendation *Rising above the Gathering Storm*. Each of these policy reform initiatives assumes the population in the United States to be a problem (lacking certain skills and literacies tied to national defense and economic growth, deficient in high-tech labor capacities compared to other advanced capitalist countries, etc.) and, as such, a necessary site of intervention to be mediated through governmental strategies, institutional management techniques, and other coercive fields of practice. One of the starkest neoliberal responses of late in the field of educational reform has been the manner in which the perceived economic and national defense crises have been grafted onto science education reform. Here, as I examine in chapter 3 in greater detail, science education, as articulated in what I call the neo-*Sputnik* policy and curricular movement, is based on a human capital model of education that fuses together economic and social health within the productive paradigm of biocapitalism.

The melding of educational reform through policies and strategies aimed at replenishing reserves of high-tech human capital also has another important biopolitical quality I am interested in fleshing out throughout the book, in particular, how neoliberal forms of educational governmentality have to an even greater extent compressed national security concerns with national education goals. Viewed from the perspective of national security, the educational population under neo-*Sputnik* strategies of reform is framed and understood within a larger global struggle for market and financial domination in an increasingly "flat world" where the United States is perceived to be in steady decline. Mantras such as innovation, competitive advantage, and "value added" have become key pedagogical values trumpeted by politicians and corporate stakeholders (not necessarily mutually exclusive) in their legislative responses to bolstering sectors of the global economy that promise the most room for growth such as the biotechnological and biomedical industries. Racing to the top, in other words, has more to do with outcompeting nations such as Germany, China, India, Korea, and Singapore in high-tech areas of the economy that require specific types of human capital than it does with good test scores, those these are one main tool of human capital measurement. Governmentality within the context of educational reform and practice in the neo-*Sputnik* era should, therefore, not solely be understood as a domestic project; it also operates within broader networks of biopower that seek to manage and control life to fit into a particular global economic reality. Such a reality, as I argue in the following section, is built on an educational foundation that can best be described as one concerned with producing optimal types of human capital built on modes of behavior, skills, literacies, and habits that underpin what Hardt and Negri have identified as the productive base of late capitalist society: immaterial labor. As such, understanding neoliberal governmentality strategies of education in the "flat world" requires an analysis of the skills, behaviors, and ways of thinking and communicating that fit within the productive universe of biocapitalism. I am also turning to Hardt and Negri's concept of immaterial labor because it underscores how, under global neoliberal forms of governmentality, new qualities from labouring bodies are required and thus offers a better understanding of the kind of educational human capital biocapitalist growth demands and the networks of power involved in its production.

Immaterial Labor and Empire: Learning the Productive Base of Biocapitalism

Hardt and Negri's concept of immaterial labor cannot be understood outside their theory of Empire. As such it is necessary to enumerate some of the key attributes of Empire in their work, and specifically how it extends forms of governmentality performed by the state originally uncovered by Foucault. Hardt and Negri's concept of Empire, in many ways, can be thought of as an analysis that charts sovereign power's transition from the state to the diffuse circuits of global capitalism. Yet Empire exhibits a number of key differences from modern critiques of imperialism usually situated within the power dynamics of the nation-state and colony as well as Foucault's study of biopower that largely rested on the state.[7] Among the most important of these differences in how biopower is articulated within Hardt and Negri's (2000) construct of Empire are the following: the lack of territorial boundaries (both in the spatial and biological sense); practices and expressions of sovereign power that operate under the appearance of a timeless reality or natural state of things; cloaking market and military interventions with ideologies of peace and democratic development; and the establishment of institutions and social mechanisms of control to regulate the productive capacities of subjectivities within a population or territory. Through transnational corporations such as Monsanto or Halliburton, and institutions such as the IMF, World Bank, NGOs, and the United Nations, Hardt and Negri (2000) argue that biopower's reach has become diffuse in nature through these networks of global capital and governance that "not only manages a territory and a population but also creates the very world it inhabits. It not only regulates human interactions but also seeks directly to rule over human nature. The object of its rule is social life in its entirety, and thus Empire presents the paradigmatic form of biopower" (xv).

Hardt and Negri's understanding of biopolitical production in the context of Empire also offers another key analytic component that is helpful for understanding the ways in which educational subjectivities are being shaped through types of educational governmentality in the biocapitalist era. That is, existing at the productive base of biocapitalism, the types of human (and nonhuman) creativity and work taking place in the laboratory and marketing firms of big pharma, for example, are what Hardt and Negri have called immaterial labor in their work. The relationship emerging between education and forms of immaterial labor is important to excavate because at both the material and symbolic levels, biocapitalism requires and advances the primary mode of labor that Hardt and Negri argue is the ontological foundation for the biopolitical production of subjects within Empire. In this sense, Hardt and Negri's dialectical critique of immaterial labor is important to my analysis of education within the biocapitalist era because it helps discern how the productive needs of biocapitalism intersect with the educational subject of the future and the type of investments being made into the processes of schooling along these lines. Specifically, immaterial labor, as the dominant model of socially necessary labor in the late capitalist epoch, represents in general a desired set of productive capacities that biocapitalist modalities of education is seeking to cultivate within the educational population of the United States. In addition, by understanding immaterial labor as one of the primary pedagogical goals of neo-*Sputnik* reform, it is easier to see how education is becoming more deeply connected to economic and security projects that are highly invested in projects pushing unrestricted economic growth and global hegemony in areas of science and technology, military/security apparatuses, and resource acquisition (including forms of human capital).

As Hardt and Negri have argued in their mapping of Empire and modes of resistance that can emerge from within its productive spaces, immaterial labor is increasingly becoming the dominant form of labor in most advanced capitalist nations. Immaterial labor, however, differs from industrial modes of labor in that "it creates immaterial products, such as knowledge, information, communication, a relationship, or an emotional response" that can be broken down into two principle forms according to Hardt and Negri: intellectual or linguistic and affective labor. Intellectual or linguistic labor is characterized by "problem solving, symbolic and analytic tasks, and linguistic expressions. This kind of immaterial labor produces ideas, symbols, codes, texts, linguistic figures, images, and other such products" (Hardt and Negri 2004, 108). The second type of immaterial labor has to do with kinds of labor that require "affects such as feeling of ease, well-being, satisfaction, excitement, or passion" (Hardt and Negri 2004, 108). What Hardt and Negri mean here are fields of work that are built on service and affective gestures in areas such as nursing, customer service and public relations departments of corporations, or teaching, for example. For Hardt and Negri, immaterial labor is also a key concept in their theory of Empire and associated model of resistance, multitudinal democracy. Their analysis suggests that immaterial labor is becoming the unifying creative force enveloping the social lives of humans and nonhumans within the expanding circuits of global capitalism and, as such, is also the common ground on which forms of resistance can emerge. In other words, if Marx's proletariat class was the revolutionary force of the industrial age due to the centralizing powers of the capitalist mode of production, immaterial labor for Hardt and Negri is following a similar path in that immaterial labor, as a social mode of productive human creativity, is the basis for current and future forms of resistance, though in a less uniform manner from what Marx and Engels envisioned.

Hardt and Negri's conceptualization of immaterial labor and their claim that it is becoming the hegemonic form of human work across the globe has also been met with a good deal of criticism (Dyer-Witherford 2001, Zizek 2001, Borón 2005, Camfield 2007). As their critics have asserted, much of the world is far from being a neo-proletariat mass congealed through immaterial forms of labor. Indeed, much of the productive base of capital still resides in the global South and Asia which resembles industrial types of work more than immaterial forms (maquiladoras and the "sweat shops" of Southeast Asia, for example). Despite Hardt and Negri's (2004) recognition of this fact, they nonetheless argue that since immaterial labor is ever more important to modes of work in late capitalist societies, it will eventually subsume all productive spaces across the globe. While their overall framing of labor or the harnessing of human forces of creativity and imagination through biopolitical modes of production may exaggerate immaterial labor's universality, it nonetheless provides a particularly well-suited framework for understanding one important way transnational corporations, governments, and venture capital are shaping education in the biocapitalist era.[8] Indeed, it is within an educational terrain where the skills and knowledge practices are used in the laboratories of Monsanto, Merck, Dow Chemical, and AquaBounty Technologies Inc., all industries based on immaterial forms of labor, where the desired goals of the educational future in the United States are playing an increasingly important role in setting the tone for the classroom.

One setting that exemplifies the centrality of immaterial labor to biocapitalist production is the genetic food industry. As chapter 4 analyzes in greater detail, the mushrooming genetic food industry is on the cusp of a historic breakthrough: the ability to produce and market a form of commodified life created through a mixture of genetic material from two different species of fish. Perhaps the most famous case since Dolly's appearance in 1996, AquaBounty Technologies Inc.'s rewriting of the FDA regulatory rules that will allow them to introduce a "new animal product" into the food system is paradigmatic of a biocapitalist developmental strategy. What AquaBounty's GE salmon (named the AquAdvantage salmon) represents, among other things, is the first animal food product made for human consumption by the aquafarming industry which is considered to be one of the fastest growing and most profitable sectors of the bioeconomy.[9] Yet at the heart of this productive animal lies a perfect example of immaterial labor's importance to biocapitalist growth and the extent to which it is involved in reassembling and producing novel forms of life to be sold and invested in as the food commodity of the future.

The knowledge and technologies that allowed the AquAdvantage salmon to be brought to life as the first flesh commodity rest on bioengineers' ability to manufacture an rDNA construct, something that involved the splicing of two disparate species of fish DNA: an Ocean pout and Chinook (king) salmon. Put differently, it was the deciphering of genetic codes; the interpretation and reconfiguration of such codes and information; the skill and knowhow to work with complex images, constructs, and models; and the linguistic aptitude of computer programmers to write the software that allowed such a scientific practice to occur in the first place. However, there was another equally important tool used to bring the AquAdvantage salmon to life. This was the giant affective apparatus built to convince the public and lawmakers that AquaBounty Technologies' laboratory fish is safe to eat and even offers a model of "green" food production. Biotech experts, ecological scientists, and advertising firms were all mobilized in an effort to push this finned invention up a stream of doubt and public uncertainty. The resources put into this affective labor no doubt were as costly as the material production that took place in the laboratory to birth the GE fish from an rDNA code to the fastest growing salmon on earth. It is in modes of immaterial labor in the laboratories of AquaBounty Technologies and the public relations extensions that are helping normalize how life will be both thought of and utilized in the classroom: as a manageable and desirable site of manipulation for economic growth.

Hardt and Negri's theory of immaterial labor, in addition to helping elucidate the co-productive relationship developing between biocapitalism and education, also demonstrates how schooling in the biocapitalist era is enmeshed within a global battle to be a world leader in human capital production invested in immaterial labor. There is both an internal and an external dimension to the US's approach in responding to the need for greater reserves of human capital invested with immaterial labor literacies that is important to recognize. From an external perspective, one way US policy makers and corporate stakeholders are reacting to the sharp need for more immaterial labor, especially in the STEM areas, is by adjusting educational reform to the perceived "flat world" geography in which biocapitalism as an economic model is unfolding. For example, the

Institute of Bioengineering and Nanotechnology (IBN) located in Singapore is designed to be a leading international site of research and development and training facility for the biomedical industry. IBN represents one of a growing field of biocapitalist institutions that are actively seeking "to position Singapore at the cutting edge of global biomedical research" that is interested in "bridging the gap from bench to bedside" (IBN 2011). As a site of valuable human capital steeped in the immaterial labor desired by biocapitalist entities, the IBN represents to corporate leaders and policy makers a simultaneous threat and model to emulate in our neo-*Sputnik* moment.

Given that Singapore is a leading country in the area of biomedical research and development, IBN reflects precisely the competitive biocapitalist culture in which current educational reform strategies in the United States are being framed in public debate. Such a highly competitive context has led policy makers and industry leaders in the United States to forefront the lack of a domestically trained labor force needed by the biotech and biomedical fields as central to the connected crises facing education and economy in the United States. The entwined educational and economic crises, according to one of the most influential education policy analysts in the United States, Thomas Friedman, have forced transnational biopharmaceutical and biomedical corporations to farm human capital reserves overseas in order to fuel growth in a highly volatile and quickly changing knowledge-intense field.[10] It also creates the internal (domestic) drive by corporate lobbyists and politicians to produce governmental strategies for addressing the deficit in human capital needed by this powerful sector of the economy. To see such an example of governmentality oriented toward the production of immaterial labor, one need only look to President Obama's "Strategy for American Innovation" (SAI) that situates education at its foundation and represents an interior dimension of biocapitalist production geared toward addressing the immaterial labor needs of the nation.

At the base of Obama's marquee economic revitalization plan, "invest[ing] in the building blocks of American innovation" is a decidedly human capital concern. Here, the project to boost domestic production of immaterial labor in the United States is clearly evident and rests on the potential to increase human capital as a catalyst for economic innovation: "We must first ensure that our economy is given all the necessary tools for successful innovation, from investments in research and development to the human, physical, and technological capital needed to perform that research and transfer those innovations" (National Economic Council 2009). What must not be lost from sight, however, is the fact that the SAI is rooted in an aggressive ideology of global economic competition. That is, in addition to its educational dimension, the SAI also seeks to "restore American leadership in fundamental research" as well as actively "promote competitive markets that spur productive entrepreneurship" (National Economic Council 2009). In the latest update to the SAI (2011), President Obama also emphasized the importance of the private sector as the engine to innovation, pledging massive government support for public/private partnerships to power his economic revitalization project that states to "win the future, we must out-innovate, out-educate, and out-build the rest of the world" (National Economic Council 2011). What I am suggesting must be recognized in such a strategy are the types of governmentality being articulated via a neo-imperial ethic aimed at outcompeting the rest of the world, and also a *productive* one that has set its sites on investing in domestic populations in order to create surplus reserves of immaterial labor tuned to the needs of a rapidly augmenting biocapitalist economy.

Current reform strategies aimed at funding STEM education at home and importing graduate students from countries rich in scientific, mathematical, and engineering skills and know-how can, therefore, be seen as two sides of the same coin. Governmental strategies shaped by biocapitalist imperatives thus require types of intervention to manage and regulate the educational population of the United States through mechanisms that construct particular contexts for fields of practice and action, such as the funding structure of Race to the Top, SAI's focus on education for increasing immaterial labor, and by incentivizing corporate/university partnerships in order to ensure a greater output of human capital in key economic areas. In short, the merging of biocapitalist imperatives with education cannot be fully understood outside of the global networks of power where entities such as the IBN institute and domestic policies such as SAI are heavily investing in productive sites of immaterial labor, the training lab and the school being two of the most important zones of interest. What can also be discerned from these examples is how immaterial labor functions within Empire's fluid and shifting forms of biopolitical control. Different from Foucault's concept of governmentality, Hardt and Negri's reworking of the concept recasts practices of governmentality within a global network of transnational corporate competition, military/nation building projects, and pro-market legal apparatuses that facilitate types of life required to participate in the unsustainable neoliberal fantasy of achieving 3 percent compound growth in GDP per year without end (Harvey 2010). Where biocapitalism fits into and fuels such a growth model is important to the way we understand how neoliberal strategies of governmentality are involving education in disturbingly novel productive habits that have cast educational life as a key actor in a global struggle to get more out of life.

Here we can begin to see how neo-*Sputnik* educational reform, understood through the economic/national security focus on immaterial labor, situates life within an extractive field of growth. Such an extractive field, one where human vitality is treated and understood as a minable value, I am suggesting, is being shaped within the terrain of neo-*Sputnik* education policies and rationalities driven by biocapitalist imperatives. The extractive field that has emerged from biocapitalist pressures on educational reform is a result of an model of economic growth that must target new forms of life in order to continue a consumptive and productive process that has already pushed well beyond its sustainable limits. The productive subjectivities that schools potentially hold as promissory reserves of human capital surplus, I am arguing, is one of the most important sites where such an extractive field of vitality has come into being. Indeed, the relationship developing between education and forms of neoliberal governmentality must be rethought within the growth model of biocapitalism that is built on an extractive ethic not unlike the one guiding biocapitalist corporations search for useful genetic information in plant and animal species to create the next breakthrough drug. The emergence of extractive schooling, in other words, is a consequence of an economic and cultural context that is increasingly geared toward capturing latent value from forms of both biological and social vitality.

The co-productive relationship between biocapitalism and education, I argue throughout the chapters of this book, needs to be read as a site of biopolitical production. Institutional partnerships and policy reform such as the SAI or the Merck Institute for Science Education are involved in projects seeking to manage and regulate both individuals and populations not solely from the exterior (structural forms of coercion) but, as Hardt and Negri have argued, perhaps more importantly, from the interior (shaping and directing human habits, values, ways of learning and teaching, and desires). In other words, in order for a biocapitalist entity such as AquaBounty Inc. to have market success with their fish, a populace agreeable and involved with systems of production and consumption needs to be created to perform and accept certain configurations of immaterial labor over others. One of the primary ways the biopolitical production of subjects in education is taking place today is through a variety of forms of governmentality enacted through a heavily weighted science and engineering focus on reform, public pedagogies associated with FDA and EPA deregulation efforts, and media campaigns from biocapitalist entities that guide individual and social conduct conducive to biocapitalist imperatives. One of the key goals of education in the biocapitalist era, therefore, is to naturalize immaterial labor as the material and creative force for capturing the producitive capacities of forms of life. Extractive approaches to school, in other words, have a pedagogical intent of teaching subjects that the use-value of immaterial labor is at its highest when applied to the problems and concerns of biocapitalist entities and, as a result, alternative arrangements of immaterial labor that are not based in an extractive relation to life seem irrational.

Given the biopolitical plane of globally networked systems of biocapitalist production based on immaterial labor, the obvious question must be asked: how can an alternative biopolitics of education based on a non-extractive ethic be reconstructed in pedagogical contexts? In Hardt and Negri's recent work theorizing multitudinal spaces of resistance, they have gestured toward zones of biopolitical resistance capable of pragmatically generating counter strategies based on what they term "biopolitical reason." Here, Hardt and Negri (2009) look to practices that have emerged within political and social movements such as various expressions of antiglobalization politics that work from a "rationality [that is at] the service of life; technique at the service of ecological needs, where ecological we mean not simply the preservation of nature but the development and reproduction of 'social relations', as Viveiros de Castro says, between humans and nonhumans; and the accumulation of wealth at the service of the common" (125).[11] If the political stake of biopolitical production in education rooted in forms of immaterial labor is life itself, what I suggest throughout this book is that resistance and modes of alternative life must also begin from the very same location. Identifying such practices and sites of biopolitical resistance in educational settings is one of the central goals this book takes up. Yet to start rethinking education from the standpoint of "biopolitical reason," it is first necessary to examine more fully the ways in which biocapitalism and education are coevolving within the current constellation of neoliberal restructuring projects.

Biocapital: Toward a Political Economy of Educational Life

This book is interested in understanding how biocapitalism as an economic and financial regime and technoscientific practice that reconfigures forms of life into commodities shapes and influences extractive models of schooling. Studies on the emergence of forms of biocapital have been developed over the past 15 years largely from the work of scholars in the areas of science and technology studies, anthropology of science, feminist theory, and sociology of science. This transdisciplinary field comprising the area of biocapitalist studies takes as its point of departure the economic and technoscientific transformations that have accompanied the explosion of work in genetic and molecular sciences over the past two decades. Mapping the rapid cultural and economic changes brought on by new practices in the genetic sciences that have made rDNA constructs (the cellular basis for genetic engineering), proteomics (the study of proteins), or genetic testing during human pregnancy a part of everyday life in advanced capitalist nations, theorists of biocapitalism comprise a unique tier in biopolitical studies. Whereas biopolitics articulated through the work of Foucault and Hardt and Negri focus on the production of subjectivities and technologies of control that govern human populations, theorists of biocapital focus on the ways new scientific practices have allowed for more biological material to become integral to the circuits of capitalist production and exchange. In other words, if biopolitical critique in the work of Foucault and Hardt and Negri look at the manner in which the social and political life of humans are produced through the lens of biopower, theorists of biocapital expand biopolitical analysis into the life of the nonhuman and the cellular domain of humans.

As Stefan Helmreich (2008) has noted, biocapital is a term that "paging back to Marx, fixes attention on the dynamics of labour and commoditization that characterize the making and marketing of such entities as industrial and pharmaceutical bioproducts." Yet it also

> extends Foucault's concept of *biopolitics*, that practice of governance that brought 'life and its mechanisms into the realm of explicit calculations' (Foucault 1978, 143). Theorists of biocapital posit that such calculations no longer organize only state, national, or colonial governance, but also increasingly format economic enterprises that take as their object the creation, from biotic material and information, of value, markets, wealth, and profit. The biological entities that inhabit this landscape are also no longer only individuals and populations—the twin poles of Foucault's biopower—but also cells, molecules, genomes, and genes (ibid., 464).[12]

Within the biocapitalist studies literature, two primary categories of focus can be rendered (Helmreich 2008). The first group of theorists, of which Kaushik Sunder Rajan and Malinda Cooper's work exemplifies, focuses on biocapitalist actors such as genetic start-up labs and the co-constructive relationship they retain with governmental and NGO actors. Cooper and Sunder Rajan's analysis of the political economy of biocapitalism and the subjectivities that inhabit biocapitalist sites of production has uncovered a neoliberal ethic that is animated by a messianic, pseudo-religious/entrepreneurial belief in the market combined with venture capital gambles to bring the next big biotech product to consumers. Sunder Rajan, for example, offers in his ethnographic account of genetic start-ups a picture of the neo-Weberian subjectivity that melds religious-like beliefs of salvation with biotechnological progress. Cooper's work, complimenting Sunder Rajan's, looks at how governmental strategies designed around speculative growth models drive biocapitalist production by constantly seeking and appropriating new biological materials and turning them into commodifiable goods (Sunder Rajan 2006, Cooper 2008).[13] In other words, one of the primary lines of enquiry both theorists examine is how biocapitalist entities utilize security and trade apparatuses such as the UN, WB, NAFTA, or the Trade Related Aspects of International Property Rights (TRIPS) agreement to channel the benefits of drug research and development from developing world contexts into advanced capitalist nations, and specifically the productive engines of the biopharmaceutical and bioscience industries. For this area of biocapitalist studies, the synergy that has formed around scientific activity (especially the life sciences) and governmental policies and practices oriented toward the dissolution of biological and market barriers to growth points to a unique phase in the development of capitalism.

However, as Sunder Rajan (2006) makes clear,

> biocapitalism does not signify a distinct epochal phase of capitalism that leaves behind or radically ruptures capitalism as we have known it... Rather, the relationship between "capitalism" (itself not a unitary category) and what I call biocapital is one where the latter is, simultaneously a continuation of, an evolution of, a subset of, and a form distinct from the former. Further, biocapital itself takes shape in incongruent fashion across the multiple sites of its global emergence. (10)

Yet what makes the union between science and neoliberal forms of governmentality constituting biocapitalism particularly dangerous are the ways in which biological life has become integral to economic growth in ways never before imagined. In mapping this productive trajectory of biocapitalism, Cooper (2008) notes that "the two sides of the capitalist delirium—the drive to push beyond limits and the need to reimpose them, in the form of scarcity—must be understood as mutually constitutive." As a result, Cooper argues that "the tensions of capitalism are being played out on a global, biospheric scale and thus implicate the future of life on earth" (49).

The capitalist "delirium" Cooper identifies as being at the heart of neoliberal/life sciences' productive projects (i.e., GE food or new drug discoveries) is one that also affects how education is being mobilized as a material source to feed such a frenzied state of mind.[14] Expanding Cooper's analysis of biocapitalism into the field of education, I suggest that the delirium fueled by a promissory vision undergirding biocapitalist growth allows us to see how the valorization process inherent in the capitalist system of production is being extended into the debt field of the educational population in the United States. In hedging the future of economic growth on the hope that a surplus of human capital steeped in immaterial labor can be extracted from educational zones of production, schooling in the biocapitalist era has been enmeshed within the growth delirium Cooper indentifies in her analysis of biocapitalism. Thus, biocapitalist theorists such as Sunder Rajan and Cooper provide a powerful framework for understanding the ways in which education (both universities and K-12) are involved in a political economy increasingly guided by an ethos of "futuricity" that has set into motion an insatiable search for the productive potential of more and more forms of life. Educational vitality understood through the lens and practices of human capital strategies, as I argue in chapters 1 and 2, is one such form of life biocapitalism is seeking to absorb into its metabolic system. Getting the most from educational life, in other words, is crucial to biocapitalist development as its perceived future rests on the promise of capturing latent value from an educational population invested with literacies and skills to drive economic development for yet unrealized productive futures.

I am adopting from this arm of biocapital studies their analytic focus on how subjectivities tied to what Sunder Rajan calls a "salvationary" ethic are produced within a neoliberal economic model based on "futuricity" or promissory valuation. Specifically, one of the major threads running throughout this book is an examination of the ways educational discourse and policy debate in the neo-*Sputnik* era are in large part driven by future oriented claims and a messianic belief in the power of biotechnological and biomedical industries to pull the Unites States out of economic stagnation and competitive malaise. Thus, one of the central claims I make in this book is that in order to better understand the current neo-*Sputnik* fervor propelling public discourse on education reform and, in particular, the curricular tunnel vision on STEM areas, each must be understood within the speculative context that biocapitalism generates. Such a reform environment, I contend throughout this book, is being shaped by a type of "casino capitalism" where "commercial realization, as in any other capitalist enterprise, involves a successful venture that provides a high return on investment, increased revenue earnings, and corporate growth—value systems whose ideologies are increasingly globalizing and homogenizing", one that is steeped in an "ideology of innovation—a high risk, free market frontier ideology that is simultaneously particularly American and globalizing" (Sunder Rajan 2006, 113). As I explore in multiple chapters of this book, it is in such a promissory terrain that education in the United States must now be situated and understood, as well as any alternatives to the powerful forces that have targeted both educational life and biological life generally as extractable resources to be utilized for economic growth and national security in a flat world.

The second area of biocapitalist studies that I draw upon in this book is derived from the fields of feminist science studies and sociology of science. This branch has centered on the dramatic technoscientific advances that have transformed how we understand the human body, reproductive processes, and racial categories, and has led to what Nikolas Rose has termed the domain of "somatic ethics." Building off Donna Haraway's pathbreaking research on the ways nature and culture have been blended into entirely new hybrid forms of life through technoscientific practices in corporate laboratories, this branch of biocapitalist studies is largely concerned with the question of how the biological has become an isolatable field of potential profit and intervention through bioscientific research and practice (Haraway 1990, 1997, 2008). As Sarah Franklin and Margaret Lock (2003) have put it more recently, "biocapital is driven by a form of extraction that involves isolating and mobilizing the primary reproductive agency of specific body parts, particularly cells, in a manner not dissimilar to that by which, as Marx described it, soil played the 'principal' role in agriculture" (8). What Franklin and Lock, as well as Charis Thompson (2007) in her work looking at reproductive technologies, have argued is that bioscientific interventions into biological processes such as human or cellular reproduction have opened up a whole new *productive* field for capital based on the ability to capture and redirect the reproductive capacities of organisms into market circuits. Working from the standpoint of kinship studies, feminist theorists of biocapitalism have powerfully demonstrated how not only life itself is a commodity but also the biological processes that regenerate and substantiate new forms of commodified life have been enlisted into novel productive regimes (Franklin and Lock 2003).[15]

One of the most instructive areas to see how biocapitalism sustains its productive base through capturing value from the reproductive capacities of organisms is in what has been termed "bioprospecting." As Cori Hayden (2003) defines it, bioprospecting is "the new name for an old practice: it refers to corporate drug development based on medicinal plants, traditional knowledge, and microbes culled from the 'biodiversity-rich' regions of the globe—most of which reside in the so-called developing nations" (1). In Hayden's work that looks at the network of relations involved in the extraction of biological materials from plants in densely biodiverse areas of the earth such as Mexico, a pathway from the jungle to the latest biopharmaceutical breakthrough drug on the market is shown to consist of a whole host of institutional, legal, local populations, and scientific actors. Here it will be helpful to look at an example of how the practice of bioprospecting involves multiple levels of life in a powerful productive network. Specifically, it will be useful to look at how one of the most profitable pharmaceutical products in modern history was brought into existence from a gnarled tuber found in the jungle highlands of Oaxaca, Mexico, to regulate the reproductive capacities of women's bodies.

One of the largest bioprospecting enterprises of the twentieth century spawned out of the pharmaceutical industry's discovery of a hormone unique to the barbasco plant. Embedded in the story of how a wild yam in Southern Mexico was transformed into a biochemical value by US and European scientists is an important aspect to biocapitalism: scientists' ability to recognize extractable value lying dormant in the reproductive capacities of organisms (their biochemical capacities for growth). As the cornerstone to the oral contraceptive industry, the barbasco plant itself held the biochemical key to reproducing synthetic chemical compounds that would alter the drug market and the political landscape of human reproduction henceforth. As Gabriela Soto Laveaga (2009) reveals in her excellent history of the barbasco plant and "the pill," "from 1940 to the mid 1970's, these yams were the ideal source material for the global production of synthetic steroid hormones. Mass production of progesterone, cortisone, and eventually, oral contraceptives was possible because of the availability of Mexican yams" (2). Yet the barbasco plant did not only signify an instance where science was able to harness the reproductive biochemical processes of a plant, but it also marks the advent of a powerful biotechnology that has intervened into the reproductive capacities of women's bodies.

The history of the barbasco plant and the processes involved in bioprospecting its latent value by the pharmaceutical industry and state governments provides at least three important features of biocapitalism worth considering. The first involves the plant itself, the biological material necessary for creating a commodity for the pharmaceutical market that is largely driven by a highly speculative venture capital model and ecologically and culturally imperial set of practices. That is, as Laveaga's (2009) history of the barbasco industry describes in great detail, it took massive governmental and private coordination as well as venture capital to build the industrial infrastructure in rural Southern Mexico that allowed the harvesting, processing, and shipment of pharmaceutical products across the world. Next, the physical extraction of the plant in an incredibly unforgiving terrain required a massive amount of human labor and local knowledge of the plant (where it grows, how to harvest the tuber without damaging the root system of the plant, etc.) to be organized and exploited. As Hayden (2003) notes, bioprospecting thus also requires the conditioning of subjectivities all along the line of production; to the peasant laborers on the side of the mountain to the laboratory workers and the venture capitalist investing in start-up drug companies and university partnerships, there are multiple subjectivities at play in this chain of relation that moves from mountainside extraction to corner pharmacy.

Finally, the example of the barbasco plant directly implicates biocapitalist practices into reproductive debates and the lives of women in society. In other words, while the production of the synthetic steroid needed to create "the pill," on the one hand, offered to women a new degree of control of their reproductive lives, on the other, it allowed white male scientists and expert knowledge once again to intensify its reach into women's bodies. As Michelle Stanworth (1987) has argued, "medical and scientific advances in the sphere of reproduction—so often hailed as the liberators of twentieth century women—have, in fact, been a double edged sword. On the one hand, they have offered women a greater technical possibility to decide if, when and under what conditions to have children; on the other, the domination of so much reproductive technology by the medical profession and by the state has enabled others to have an even greater capacity to exert control over women's lives" (11).

As the example of the barbasco plant suggests, enmeshed in the productive paradigm of biocapitalism is a propensity to deeply involve the processes of life with the circuits of capital exchange and production—in fact, its developmental growth is based on the technoscientific ability to intervene and extract value from organism's regenerative and reproductive capacities. What this second area of biocapitalist studies, therefore, offers is a powerful tool for making sense of (1) how technoscientific practices have reached into the reproductive powers of life, opening it up to market logics, and (2) the ways in which individuals relate to their bodies in a historical moment where intervention into more and more aspects of biological life is becoming routine. On this last point, the work of Nikolas Rose that looks at the biopolitical dimensions of biomedicine in society and culture provides another important insight into features associated with the productive apparatus of biocapitalism, namely, how human health in an era of rapid biomedical and biotechnological change has lead to the emergence of individuals and groups who are increasingly understanding themselves through what Rose calls "somatic ethics," or an ethic oriented toward the care of the body. According to Rose, individuals within the biomedical age are more intensely coming to define themselves through a type of somatic ethics that is partially a result of advances in biomedical and biotechnological practices and the public pedagogies that emanate from their productive apparatuses. As Rose (2007) puts it, "we are increasingly coming to relate to ourselves as 'somatic' individuals, that is to say, as beings whose individuality is, in part at least, grounded within our fleshy, corporeal existence, and who experience, articulate, judge, and look upon ourselves in part in the language of biomedicine" (25–26). In an age where the human body can be managed at the cellular level, Rose's work points to the emergence of a reflexive relationship that has developed between the public pedagogies of the biomedical industry and the individual's understanding of his or her body. The shear range of drugs and treatments that can intervene into the processes of the body, in other words, has fundamentally changed social life according to Rose's biopolitical analysis and thus requires a rethinking of political categories of the self.

As a result of the reflexive relation developing between the human body and biomedicine, Rose sees what he calls "biological citizenship" as a constituent feature of the current political landscape of advanced industrialized nations. Biological citizenship for Rose (2007) is in large part the byproduct of biomedical and pharmaceutical companies' educational campaigns that

> set up and sponsor many of the consumer support groups that have sprung up and around disorders from attention deficit hyperactivity disorder (ADHD) to epidermolysis bullosa (EB). In doing so, they seek to represent their activities and their products as beneficial, to counter the claims critics, and to educate actual or potential consumers of their products. In the United States pharmaceutical companies are permitted to engage in "direct to consumer advertising" and television advertisements for the benefits of different brands of drugs are widespread: notably drugs for treating experiences of mental malaise, now coded as depression, anxiety, and panic disorders. (142)

As I argue in chapter 5, the notion of biological citizenship can also be extended from the public sphere's perceptions of the biomedicalization of the human body directly into the classroom. That is, through powerful mechanisms wielded by the biomedical industry and big Pharma such as advertising, media campaigns, and direct pedagogical intervention through counseling and expert-run workshops in schools, biocapitalist projects also encompass strategies aimed at cultivating somatic ethics in line with the rationalities associated with biomedical products and the markets they seek to enter and cultivate (Rose 2007). Indeed for companies such as Novartis who make Ritalin, the stakes are high for developing types of public pedagogies that can reassure and teach individuals that bodily care and maintenance are synonymous with biochemical interventions into the body and to teach people that their products offer the normalized solution to socially "abnormal" ailments such as ADHD. This fact has not escaped corporate leaders of pharmaceutical companies who are among the top lobby spenders yearly to make sure regulations are favorable to market growth. In this case, market growth requires access to more and more forms of human life in order to increase profit margins and maintain industry expansion. It is thus not surprising that Novartis plays a pivotal role in creating the conditions in which their products can enter and become a normalized practice of bodily control within schools, a theme chapter 5 examines closely.

Interventions from biocapitalist entities such as Novartis into schools, something that has become a normalized practice, makes a lot of sense from a market growth standpoint: schools represent an extremely profitable market and a site where corporations can construct elaborate and expensive pedagogical techniques that can lead to a model of biological citizenship that is self-referential to biomedical rationalities and the remedies they purport to offer. Attention and behavioral problems, to take one example, are filtered through expert rationalities and products made by Novartis that in turn create a type of "biosociality" within schools that directly benefits pharmaceutical corporations while naturalizing technologies of control such as Ritalin as a response to children's behavior and academic performance. Regulating for efficiency is becoming an increasingly important desired effect given the high-stakes testing culture of schools in the United States, especially in a reform climate where teachers' jobs are now on the line for their students' academic achievement. The growth model of biocapitalism thus can be seen also as retaining public pedagogies that coerce individuals and groups in such a way as to normalize the practice of medicating for behavioral control through biopharmaceutical knowledge regimes and products. In this sense, another important claim that I make in this book is that the creation of biosocialities in schools achieved through biomedical intervention can also be seen as a means for regulating the reproductive capacities of education in such a way as to optimize life within schools.

Similar to the ways biocapitalism intervenes into the reproductive processes of women's bodies, forming types of "biosocialities" that take their cue and learn from new reproductive procedures and treatments and their attendant public pedagogies (fertility technologies for example), educational life, I am suggesting, should also be understood as a reproductive site open to a variety of forms of biocapitalist intervention. For instance, the educational potential of the population in the United States thought of in terms of its reproductive capabilities, or as a "stock of human capital" as Harvard economists Claudia Goldin and Lawrence Katz (2010) argue in their wildly influential book *The Race between Education and Technology*, reveals an uneasy symmetry growing between the lab and the school. Research on new reproductive technologies such as reprogenetics, a genetic manipulation technique aimed at achieving "brave new world" genetic choices for parents, for example, shares a similar extractive logic with human capital approaches to schooling that I examine throughout this book.[16] Schooling, within the human capital framework, can quite accurately be seen as mirroring the biocapitalist practice of "capturing . . . the latent value in biological processes, a value that is simultaneously that of human health and that of economic growth" (Rose 2007, 33). Thought of in this way, part of the project of this book is to better understand the practices and discourses of extractive schooling that are built on capturing latent "biovalue" from the educational population, something I suggest is inherent to human capital theories and approaches to education. What critical educators and communities, I am suggesting, need to begin to grapple with is how life in schools within the biocapitalist era is subject to the similar logics of sorting, assessing, testing, and maximization techniques of control and regulation that are seeking to optimize other forms of life within the productive apparatuses of the bioindustries. As educational life is being redrawn along biopolitical lines through the needs and requirements of the rapidly expanding bioindustries, it is indeed time to readjust critical theories of education and practice in a similar fashion.

Education and Biopolitical Critique

Indeed, as Lemke's survey of the evolution of biopolitical literature has pointed out, biopolitical critique offers a framework for understanding educational life as an ontological zone involved in a "network of relations among power processes, knowledge practices, and modes of subjectification" (Lemke 2011, 119). That is, what the analytics of biopolitics provides is a political epistemology that starts with life as its basis. In this sense, a biopolitical critique of education is different from traditional Marxist or critical pedagogy critiques in a number of important ways. First, traditional Marxist analyses of education largely frame the function and role of schools in society along ideological lines which, in turn, establishes a particular ontological understanding of educational life. For example, Samuel Bowles and Herbert Gintis's classic study *Schooling in Capitalist America* (1977) and Paul Willis's *Learning to Labor* (1981) situate the individual predominantly within the structural mechanisms of capital and specifically, the general need to reproduce a relatively healthy and educated working class from which surplus labor can be extracted. While critical pedagogy and Marxist critiques of education have indeed advanced since these seminal works, the central foci of analysis still largely remain on the role schools play in class formation and ideological configurations of consciousness that tend to lead to a uniform view of the body and mental conceptions of the world (Freire 2000 [1970], McLaren 2005, Allman 2001).

Now this is not to say that Marxist critiques of schooling have been ill directed or that they should be abandoned. Quite the contrary is true in relation to biopolitical critique and its development as a mode of social and cultural analysis—it is perhaps more accurate to say that biopolitical critique is a resituating of Marxian forms of analyses alongside and in conversation with post-structural approaches that amplify instead of subjugate the ways in which race, gender, sexuality, and ethnicity constitute forms of power and subjectification in society, especially in an era where the lines between culture and nature have been dissolved by advances in biotechnological and biomedical practices. In other words, biopolitics "places at the inner most core of politics that which usually lies at its limits, namely, the body and life . . . Seen this way, biopolitics again includes the excluded other of politics" (Lemke 2011, 117). The plane of materiality, to think of it differently, is refocused from the standpoint of biopolitics to take into account the ways in which capital, for instance, maintains power through the control and management of human *and* nonhuman life by playing an active role in the production of subjectivities, cultural values, and power relations under global neoliberal economic restructuring projects—a site of struggle where education figures prominently.

Second, biopolitics, at least how it is articulated in the work of Hardt and Negri, also offers a more heterogeneous model of resistance that grows out of the uneven yet totalizing forms of neoliberal governmentality operating within Empire, as opposed to traditional Marxist notions of revolutionary social and political change via party and state organization. A multitudinal arrangement of resistant politics to contemporary forms of capital and the power relations that maintain its dominance is particularly important given the varieties of racial, gender, and class hierarchies that constitute power relations in society, as well as ecological and planetary destruction that occurs not necessarily along a continuum but within historically specific spaces and cultures that developed unevenly in the colonial and neocolonial periods. Multitudinal democracy, as theorized by Hardt and Negri and a model that I draw on in this book for rethinking forms of educational resistance through alternative biopolitical practices within Empire, rests therefore on the creation of pragmatic ethics of resistance that are born within particular spaces and cultural contexts that are able to reject the biopolitical ordering of life through forms of biopower—largely in our historical moment through the advance of neoliberal strategies to further economize more and more facets of life.

Recent efforts to bridge the literatures of biopolitics and educational theory have opened up important new areas in which to consider schooling in the high-stakes, punitive, and neoliberal context of the current moment.[17] For instance, drawing on the work of Italian political theorist Georgio Agamben, a productive space has been opened up in which to think about the educational subject within an increasingly regulated complex of surveillance, measurement, and sorting technologies associated with zero tolerance practices in schools (Lewis 2006).[18] Tyson Lewis has suggested that schools in the United States can be viewed more akin to what Agamben calls the camp, where the bodies of students are treated as nonvital, biological material subjected to pedagogical forms of sovereign power that normalizes the exclusion of educational life from democratic life (Lewis 2009, 2010). From a different point of departure, others have utilized Foucault's theory of biopower and human capital to theorize the educational subject as a self-investing, entrepreneurial individual shaped through a variety of governmental regimes that promote a neoliberal ethic of self-care (Peters 2005, Simons 2006, Peters and Besley 2007). This line of analysis illuminates how student subjectivities formed within matrices of neoliberal governmentality have fallen under a "regime of economic terror and learning as investment through a variety of market based interventions" (Simons 2006).

Within educational theory, there are three categories that can be delineated which are working within the biopolitical tradition. The first, which is largely represented by the work of Tyson Lewis, focuses on understanding how forms of educational biopower regulate and control the educational body. Drawing on Agamben's studies of historical modalities of juridical sovereign power that Agamben views as creating zones of exception and inclusion in political life, Lewis examines how types of educational biopower similarly creates states of exception in pedagogical settings. Specifically, Lewis focuses on Agamben's notions of "bare life" (zoë) and political life (bios), and in particular, how they are pedagogically produced in the increasingly prison-like atmosphere of schooling in the United States.[19] Here, biopolitical analysis centers on charting ontological orderings of student subjectivities that are produced through technologies of surveillance and other post-Columbine security measures that intervene into everyday life in schools. Lewis's biopolitical framing of students' bodies is instructive in that it vividly illuminates the ways in which constant monitoring and disciplinary mechanisms can lead to a camp-like environment within schools, one where life is minimized within zones of repetitive and standardized functioning.

In response to these forms of negative or "necropedagogical" forms of biopower, Lewis has also turned to the work of Hardt and Negri to theorize how new democratic subjectivities and spaces might be pedagogically rendered by drawing upon their model of multitudinal politics that grows out of a response to Empire's expressions of biopower. In so doing, Lewis links Hardt and Negri's democratic model of the multitude with Dewey's notion of experimental learning and habit—thus, offering a biopolitical grounding in education for producing subjectivities where "Dewey's pedagogy becomes a *biopedagogy* against the various levels of corruption emanating from within Empire" (Lewis 2007, 700). While Lewis has certainly opened up new lines of investigation through his biopolitical critique of educational theory and practice, his analysis is limited in that it does not touch upon the ways in which neoliberal encroachments into educational life have transformed the productive landscape of schooling itself. That is, while Lewis has identified the terrain of bare life in education, his biopolitical critique ultimately leaves unexplored how the processes of extractive schooling molds and shapes subjectivities through types of investment and resource removal necessary for biocapitalist growth.

The second frame in education used to examine educational life from a biopolitical perspective can be broadly categorized as governmentality studies in education. This field of theorists has built on Foucault's analysis of neoliberalism and particularly how practices of national and supranational governmentality have become perhaps the most important site where state and corporate intervention is shaping educational life in a variety of ways. Thus according to Michael Peters (2005), the convergence of neoliberal economic theory with forms of governmentality creates "an intensification of moral regulation resulting from the radical withdrawal of government and the responsibilization of individuals through economics. It emerges as an actuarial form of governance that promotes an actuarial rationality through encouraging a political regime of ethical self-constitution as consumer-citizens" (131).[20] From the purview of governmentality studies in education, at least two important biopolitical areas of focus are called to our attention. The first has to do with the overall subsuming of public institutions of education under the aegis of neoliberal projects aimed at dismantling and privatizing state- and government-funded education. The second points to the formation of subjectivities under such regimes of educational biopower.

In the context of the United States, the systematic neoliberal assault on public institutions of education has expressed itself through an array of policy reform, financial deregulation, and strategic investments into "failing" or damaged public sectors of society. At the K-12 level, the effects of neoliberal economic restructuring and the contradictions that emerge from these projects can be seen in the mounting examples of longitudinal neglect many urban school districts across the United States have suffered from since the late 1970s. Ironically, this is a trend that now seeks to resolve itself through the creation of a globally competitive workforce administered through an intense effort to privatize more and more aspects of public education.[21] In just one recent example, the state of Michigan legislature in the spring of 2011 ordered the closing of half of Detroit's public schools while simultaneously increasing class size to near one hundred per class in the remaining schools (Chambers 2011). In a city that has been decimated by the globalized labor patterns of transnational corporations and the economic crisis of 2008, Detroit schools are now suffering from the dueling tensions of neoliberal economic restructuring that precisely reflect the logic of this type of educational governmentality. On the one hand, public infrastructure in the city of Detroit is being eradicated and most likely will be replaced by some for-profit enterprise, reflecting what Herbert Marcuse (2011) called the "productive destruction" ethic of advanced capitalist society.[22]

On the other hand, Michigan's budget crisis was a direct result of neoliberal financialization practices in the housing loan market that led to the implosion of the entire banking industry, causing the budgetary crises that has been rippling across city, county, state, and federal governments since the beginning of the economic collapse in 2008. This second tension present in the neoliberal restructuring of schools is built on a speculative foundation that only holds up long enough to maintain and reconstitute another round of "productive destruction." What I am suggesting the Detroit school system represents is a moment in between these two tensions within the neoliberal rearrangement of schooling taking place in the United States. The next tension will manifest as an investment and speculative model of "rebuilding" the city's schools in a way that promotes subjects who will be enlisted in the "productive destruction" ethic of a globally competitive economic system—which, contradictorily, is the same type of governmentality that produced the crisis in the first place.

At the level of higher education, neoliberal restructuring has also become the normalized way of doing business and is evidenced in the many varieties of consumer-based educational offerings (flexible course offerings, online and extension programs, evening classes, etc.) as well as the wholesale redistribution of student loan regulation to the private sector.[23] In February 2011, the Obama administration bequeathed to the student loan industry one of the biggest financial giveaways in the history of higher education funding in the United States. With a single swipe of the pen, Obama effectively handed over untold amounts of money to banks and financial institutions such as Sallie Mae by deregulating the timetable for when student loan interest can begin accruing. As a result, lending institutions will make a fortune from the now ripe for-profit interest period that has just been opened up by the Obama administration's budgetary efforts to decrease spending on public goods such as education and shift it to individuals in order to backfill the financial hole created by unfettered "casino capitalism."[24] This is a particularly dubious transfer of wealth from public hands to private firms in a time when human capital understandings frame education in public discourse and the media as indispensable for acquiring good employment during a period of severe economic crisis. What has been created through student loan industry deregulation is a type of governmentality that, on the one hand, removes public guarantees in place to protect student funding from predatory forms of lending working in tandem with a perceived social need in our "Sputnik moment" to consume more education, especially graduate degrees in the sciences and mathematics that serve as the productive basis for biocapitalist sectors of the economy. Here, we see governmentality strategies operating at two different levels that theorists of governmentality studies in education theorists have identified as key features of the neoliberal restructuring of education: the establishment of a series of economic institutions and policies that have intervened into the processes by which individuals gain access to education and, directly related to these structures, the emergence of an individual and social ethic of education as a site of entrepreneurial activity and investment that guides the conduct of individuals in neoliberal society.

The "productive destruction" ethic that I am suggesting runs through the analysis offered by theorist looking at governmentality strategies of education is indeed helpful in that it illuminates institutional practices and types of moral pedagogy, informing individuals through neoliberal arrangements of education. What is left out, however, from the governmentality perspective on education, which largely has to do with the European and Australian focus on the work in this literature, is an analysis of arguably the most potent form of governmentality operating at the interface of schooling and policy reform in the United States. That is, the primary discursive and practical arena where governmentality strategies are playing out in the US context is through biocapitalist imperatives that have created the conditions where a heavy emphasis on STEM reform and investment in university/corporate research and development partnerships has become the de facto model of educational governmentality. I argue in this book that to better understand educational forms of governmentality, at least in the US context, or that which governs "the conduct of conduct" in educational settings, needs to be situated within the variety of neo-*Sputnik* strategies that are bent on extracting particular types of vitality from the educational population in the United States.

Comprising the third rail of biopolitical studies in education is what can be called disposability models of education (Giroux 2008, 2010, 2012; Saltman 2007a; 2007b). The disposability perspective, for example, argues for "the necessity to address how neoliberalism as a pedagogical practice and a public pedagogy operating in diverse sites has succeeded in reproducing in the social order a kind of thoughtlessness, a social amnesia of sorts, that makes it possible for people to look away as an increasing number of individuals and groups are made disposable, relegated to new zones of exclusion marked by the presupposition that life is cheap, if not irrelevant, next to the needs of the marketplace and biocapital" (609). Here, Giroux as well as Kenneth Saltman have correctly identified how neoliberal logics operate in educational spaces by utilizing instances of disaster to privatize public goods and foster a corollary culture wherein life itself is understood to be valueless outside the universe of market frameworks (Saltman 2006, 2007a; Giroux 2008, 2012). Kenneth Saltman's work is particularly incisive on this point as he shows in his analysis of the Edison School how venture models of schooling should be seen as a test run for larger corporate projects to privatize public education that are just around the corner. Here, educational life can be thought of as disposable in the sense that its use-value exists as a test subject in the experimental laboratory of for-profit schooling models that follow from what Naomi Klein (2008) has termed "disaster capitalism." That is, as Saltman has astutely mapped out, a clear and distinct pattern has emerged where the shocks of financial or natural disasters in a city or state are quickly followed by neoliberal economic measures to control and reorder formerly public institutions and spaces through market-based models of development (Saltman 2007a, 2010)

While part of the project of neoliberalism has certainly been an undermining and appropriation of public and common forms of life by the forces of privatization, there is also another biopolitical aspect to this trend that has not yet been theorized, namely, how in the race to "out-educate" the rest of the world that calls for a reinvestment in education through government and corporate partnerships, the very function and role of education have been locked within a field of vision so narrow that human capital is now the default goal of schooling for both democrats and conservatives alike. What Giroux and other biopolitical critiques of education have missed, in other words, is the other side of the coin of schooling in the neoliberal era. In focusing on disposability and new zones of exclusion created through neoliberal forms of schooling, only the negative or the first shock of state and corporate dismantling of education is brought into view. The other biopolitical project at work here and what can readily be seen in the current neo-*Sputnik* rhetoric and policy reform associated with "race to the top" measures is the wholesale recasting of education as both a promissory and a risk management field that increasingly is being linked to the needs of the mushrooming bioeconomy and the riches it portends. To put it another way, what I am suggesting has not been examined within the horizon of biopolitical studies of education is the productive project taking place that is effectively an ontological reordering of life through neo-*Sputnik* reform based on investments in STEM forms of human capital.[25]

As I argue throughout this book, what has emerged from the neoliberal phoenix that first took flight in the late 1970s and continues to be the primary approach for dealing with (and manufacturing) educational crisis today is a fundamental restructuring of life in schools that cannot solely be understood through the lens of disposability but also needs to be read through the need for surplus human capital that is required to fuel biocapitalist development. Just as biopharmaceutical labs require plant material and local knowledge of biodiversity in areas such as the rain forests of Brazil to operate and extract value, so too have schools become a productive site in which life is the target of powerful productive regimes and their attendant forms of governmentality. What a biopolitical critique of schooling today must take into account, therefore, is the symmetry that has formed between the biotech lab's renderings of life into profitable forms (the AquAdvantage salmon or genetic patents) and the school as a reservoir of human capital, one of the most important ingredients to future biocapitalist growth. The postgenomic era, in other words, is fundamentally transforming the social value of education to mirror the promissory future that guides and feeds bioeconomic industries.[26]

To be sure, economic and legislative trends in recent years reflect this push to get more out of educational life. Despite the worst economic climate since the Great Depression, $10.4 billion of the American Recovery and Reinvestment Act went directly to the National Institute of Health (NIH) to fund "Biomedical Core Centers" designed to increase biomedical research and innovation in the nation. Globally, biopharmaceutical giants such as Novartis recently invested over $1 billion in the Novartis Institute of Biomedical research in Shanghai. Other biotech companies are also heavily investing in Singapore and other Asian countries that are perceived to have ready and accessible surpluses of human capital, the foundation to the competitive R&D environment that underlies biocapitalist growth. What is now becoming more apparent is that biocapitalism is not just an economic and scientific enterprise unique to genetic research sites and biomedical facilities: it is also extending practices oriented toward capitalizing on forms of life into the realm of educational life. Schooling, that is, has become a key target of the powerful productive regime of biocapitalism that is in constant search for accessible human capital reserves. Corporate interests such as the Gates Foundation and the Merck Institute for Science Education are only the leading edge of a larger K-12 reform movement being carried out through "value added" classroom measures and merit pay contracts for teachers.

Further evidence of this pattern can be seen in two of the Obama Administration's largest funding initiatives that impact education, the American Recovery and Reinvestment Act and the Race to the Top. For example, $650 million of $4.35 billion of the Race to the Top are destined for Science/Technology/Engineering/Mathematics (STEM) education development. Of the $ 1 billion allocated for K-8 education in the Recovery Act, $300 million is committed to bolstering STEM education. Furthermore, at the beginning of the 2010 school year almost all nine winning state (and the District of Columbia) applications for "Race to the Top" Phase 2 funds emphasized STEM education (Robelen 2010). Indeed, the Obama Department of Education, behind its "educate to innovate" program, has designated STEM education as having "competitive preference priority" in the "Race to the Top" evaluation process, spending $3 billion annually on STEM education alone (Cavanagh 2008).

What is becoming clear in the relationship being established between biotech corporations, education policy and curricula reform, and educational outreach programs is a picture where schools have been situated as a key extractive site within the biocapitalist production model. Schooling, in other words, is an integral part to the growth of biocapitalism and, as such, educational life is one of its most valuable extractive goods. It is in this emerging terrain where biocapitalism and schooling have entered into a co-constructive relationship that this book will offer a view into and a possible way out of this promissory future of education.

Finally, the larger focus of this book that deals with the question of humans (and nonhumans) and their co-productive relationship with apparatuses of capital is certainly not a new area of enquiry. Marx himself clearly understood the ways in which human life were being absorbed into the "metabolic" processes of capital. Perhaps one of the clearest examples of a nascent biopolitical lens in Marx's analysis of capital is his account in *Capital* volume I of the ways in which the working day had become a key space of contestation and intervention for capitalist and the state regulatory laws supporting their aims:

> Hence it is self-evident that the worker is nothing other than labour-power for the duration of his whole life, and that therefore all his disposable time is by nature and by right labour-time, to be devoted to the self-valorization of capital. Time for education, for intellectual development, for the fulfillment of social functions, for social intercourse, for the free play of the vital forces of his body and his mind, even the rest time of Sunday (and that in a country of Sabbatarians!)—what foolishness! But in its blind and measureless drive, its insatiable appetite for surplus labor, capital oversteps not only the moral but even the merely physical limits of the working day. *It usurps the time for growth, development and healthy maintenance of the body* [emphasis mine]. (Marx 1977, 375)

It is in this struggle for "the free play of the vital forces of his[/her] body and his[/her] mind" that we must now understand the stakes of education in the biocapitalist era. Yet, different from Marx's time, biocapitalism has pushed the productive frontier of capital in such a way that the vital forces of our bodies are no longer solely tied to the problematic of wage labor and the industrial productive arrangement. Instead human and nonhuman vitality have itself become a commodity, an investible future, and a promissory site in which the national and economic security of the United States depends. It is indeed a political terrain where "life itself" has become a location where the future of not only human life but also nonhuman life will be made or unmade. The question then becomes: What kind of life do we want education to be for?

PART I

Origins of Educational Biocapital

CHAPTER 1

Learning to be *Homo economicus* on the Plantation: A Brief History of Human Capital Metrics

> On the basis of this theoretical and historical analysis we can thus pick out the principles of a policy of growth which will no longer be simply indexed to the problem of the material investment of physical capital, on the one hand, and of the number of workers [on the other], but a policy of growth focused precisely on one of the things that the West can modify most easily, and that is the form of investment in human capital. And in fact we are seeing the economic policies of all the developed countries, but also their social policies, as well as their cultural and educational policies, being oriented in these terms.
>
> —Michel Foucault, 1979

Over the summer of 2011, teachers in the Ogden school district just north of Salt Lake City awoke to find a new, non-negotiated contract agreement in the mail. The District Superintendent's Office informed teachers that they would need to sign a new contract agreement that included a merit pay protocol attached to a value-added performance metric, or they would no longer have a job in the fall. The story unfolding in Ogden is becoming a frequently told one across the United States. The growth of performance measurement tools such as value-added metrics and other mathematical and statistical evaluation models that claim to isolate teachers' positive and negative effects on students' classroom achievement, has gained considerable traction through the competitive structures built into reform policies such as Race to the Top where states are rewarded for adopting accountability measures focused on increasing teacher and school performance ratings. The value-added movement, which I examine in detail in chapter 2, turns out to be just another iteration in a long line of accountability discourses and practices that have promised to "fix" endemic problems in US public schools through the application of new "revolutionary" scientific models of learning assessment. Complex historical problems of educational equity such as the growing achievement gap between white students and students of color, one of the primary aims of the No Child Left Behind (NCLB) legislation, are best solved through the development and application of newfangled techniques capable of more accurately measuring teacher and student performance.

Recent phase II funding distribution for Race to the Top reflects such a belief in new scientific measurement breakthroughs and indicates how equitable and quality education is going to be achieved under the Obama administration's reform policy that puts a shiny new wrapping on an old and broken approach. Looking at Race to the Top winners, it is abundantly clear that if states want financial support in a time of drastic educational cutbacks, they must adopt a whole host of accountability reforms including value-added measures of performance and an expansion of the charter school system. Despite the fact that federal funds for education make up only a portion of school budgets, they are nonetheless one of the most important public assurances in an economic climate where state budgets (such as Michigan's) are collapsing. Not unlike the neocolonial practices employed by the World Bank and IMF that force compliance with neoliberal developmental strategies in the developing world, states either will adopt market-based accountability measures or be starved financially while the most underserved students will continue to absorb the brunt of educational disinvestment. For example, almost all the state winners of Race to the Top funding included the following provisions in their strategic reform plan: an expansion of the charter school system, a strong emphasis on STEM curriculum and programs, and the institutionalization of value-added accountability measures that focuses on teacher performance in relation to student achievement (United States Department of Education 2012).[1] Across the country from Utah, Florida, Ohio, California, Washington DC, Tennessee, Michigan, Hawaii, Texas, Wisconsin, New York, Colorado, to Georgia, performance-based assessment models such as value-added techniques are quickly becoming one of the primary ways schools will either enter the race or be disqualified at the starting line.

Yet, in its rapid ascension to being the chief strategy for the Department of Education's signature reform plan under Obama and Arne Duncan, there has been almost no discussion on or analysis of the origins of value-added metrics beyond the controversy of whether or not teacher performance ratings are an appropriate measure of student performance and academic achievement. Where, in other words, did such an influential measurement device for determining educational quality come from? In surveying the public debate swirling around value-added applications to school contexts, it is clear that this question has been entirely lost or conveniently disguised in the decidedly neoliberal context that currently shapes the educational reform debate in the United States. It also seems next to impossible to find a competing model that either of the two dominant political parties promotes—demonstrating another interesting feature of the value-added framework: neoliberal strategies for educational restructuring (such as its predecessors *Nation at Risk* and NCLB) operates from a one-dimensional politics that treats educational life as a calculable and regulatory field of economic control and extractive value.[2]

One of the most troubling aspects emerging from the neoliberal restructuring of education over the past 30 years in the United States is the way in which reform movements based on educational measurement techniques such as the value-added framework claim to have isolated the true measure of learning into what teachers can either add or subtract from their pupils. However, the design and application of such assessment models are not a new breakthrough in educational assessment research, but rather part of a larger political project that seeks to reduce complex sociocultural settings such as schools into an economic grid of value assessment and optimization. From such a context, understandings of educational environments and best practices of teaching are organized and controlled through a rationality infused with free-market strategies and a fundamentalist view of students as self-interested economic actors situated in a highly competitive "flat" world. The primary goal of this chapter is to expose the roots of the free-market fantasy advanced through value-added proponents by situating its development within the larger genealogical arc of the Chicago School of Economics free-market laboratory, in particular, their construct of human capital theory in one of their landmark studies *Time on the Cross* (TC). Starting out as a fringe economic theory focused on discovering where and how economic value is produced in different historical and social settings, human capital theory has achieved nearly universal acceptance as the practical and ethical framework in which to view and judge educational health in the United States. The human capital project embedded in neoliberal understandings of education, such as the one I examine in Robert Fogel and Stanley Engerman's *TC*, is also the birthplace of contemporary assessment tools such as value-added metrics and, as such, I argue we must understand the historical development of human capital theory as part of the broader mission of neoliberal policies and governing practices in order to better see what is at stake in accepting measurement technologics of human value as predictors of a healthy and successful educational life.

It is particularly important to critically evaluate the history of human capital theory in our current historical moment given that in the past few years, the public has been flooded with wildly popular and influential books that wholeheartedly promote human capital frameworks. Books such as Harvard economists Claudia Goldin and Larry Katz's (2010) *The Race between Technology and Education*, Linda Darling-Hammond's (2010) *The Flat World and Education*, and *New York Times* editorialist (and part-time educational policy expert) Thomas Friedman's seemingly ubiquitous "flat-world" analyses are all prominent parts of the human capital constellation in educational research and popular media.[3] Each of these authors and many more like them that populate the educational reform industry have helped normalize human capital narratives and understandings of education so much so that it has become the language everyone must speak when offering their solutions for educational reform. Such a consensus is representative of the degree to which human capital theory has become the naturalized way to talk about and frame problems such as educational inequity, economic growth and stagnation, and the role US schools play in national security challenges in the twenty-first century that have become inextricably linked to the unrealizable dream of unbounded global capitalist growth.

Given the complete collapse of economic imperatives into national educational goals, represented in books and policy reccomendations such as these and countless others, it is now more important than ever to go back and examine how such a one-dimensional cultural logic of education has been established and helped set the stage for governmental strategies aimed at the total economization of educational life. In this chapter and the next, I look at two periods in the development of human capital theory in order to shed light on the role it has played in merging economic and educational value into an indistinguishable unity. The story of human capital theory, as I discuss in this chapter, began as a distinct project that evolved out of the radical free-market womb of the Chicago School of Economics. As such, human capital theory needs to be understood within the larger context of neoliberal thought and practice in order to see how it has become integral to the success of the neoliberal restructuring of education. To do so, it will be helpful to begin with the man by which all life is measured.

Situating Human Capital Theory within the Neoliberal Project: From Homo sapiens to *Homo economicus*

Neoliberalism, as a school of economic thought and policy development, started in the United States with a small and dedicated group of economists and policy makers at the University of Chicago during the period immediately following World War II. Their mission as an intellectual community was to establish research and social policy approaches infused with the belief that a utopian capitalist state could be achieved through a radical free-market restructuring of the growing welfare state of the New Deal era. As Naomi Klein (2007) puts it, Milton Friedman and other Chicago School disciples believed in a view of pure capitalism that held "just as ecosystems self-regulate, keeping themselves in balance, the market, left to its own devices, would create just the right number of products at precisely the right prices, produced by workers at just the right wages to buy those products—an Eden of plentiful employment, boundless creativity and zero inflation" (50). The neoliberal theorists' dream of a pure capitalist state was built through the radicalization of the ideas of classic political economists such as Adam Smith, David Ricardo, John Locke, Thomas Malthus, and John Stuart Mill (Foucault 2008, Hardt and Negri 2000, Harvey 2005, Giroux 2008, Peters 2009). In one sense, neoliberalism then can be seen as aggressively retooling classical liberal economic concepts and notions of the market and state to match their goal of attaining a capitalist Garden of Eden. The primary target of the Chicago School's "new" liberalism was the state's control and administrative power over social welfare programs (including schools), international development policies, and regulatory mechanisms of commerce that hindered unfettered corporate expansion (Harvey 2005). Or, as Michel Foucault put it in his 1978 lectures on the development of neoliberalism in the United States, the overarching target for the Chicago School and the powerful supporters of its views consisted of "three elements—Keynsian policy, social pacts of war, and the growth of the federal administration through economic and social programs—together formed the adversary and target of neoliberal thought, that which it was constructed against or which it opposed in order to form itself and develop" (Foucault 2008, 217).

Within the historical development of neoliberalism as an intellectual movement seeking to remove the regulatory and legal parameters of the Welfare state that encumbered privatized growth in society, human capital theory played an instrumental role. Specifically, human capital theory serves as an important conceptual bridge that opens up more spaces to privatization such as human beings' productive life and public goods such as schools. Yet in order for more public and governmental entities and people to be transformed through the processes of privatization, market responsiveness, and value maximization, neoliberal theorists recognized that a theory and set of practices needed to be invented in order to establish the basis for human subjectivity through their developing economic science. All of human life for neoliberal theorists, in other words, needed to be evaluated in terms of human capital accumulation and investment. Put differently, if the structural dimension of neoliberalism focused on the restructuring of state institutions and legal apparatuses through free-market governing strategies and policies, a concomitant theory of the subject was also required in order to understand Adam and Eve in their true light as entrepreneurial apple farmers. However, for the ontological condition of humanity to be transformed into what Foucault, referring to Adam Smith and Jeremy Bentham's moral theory of the liberal individual, called *Homo economicus*, or, the self-interested, rational economic subject, human capital theorists such as Theodore Schultz, Robert Fogel, Jacob Mincer, Milton Friedman, Gary Becker, and other Chicago School economists had to overcome two important problematics. Standing in the way of their vision of a free and open society guided by unbridled capitalism was the question of how to ontologically establish human nature and social environments seemingly inseparable from free-market rationalities of the pure capitalist state.[4]

The first problem Chicago School theorists confronted while trying to conjure the neoliberal subject into existence was the issue of living and dead labor. That is, neoliberal theorists had to grapple with the question of how corporations, governments, and individuals in capitalist nations and the developing world could be recalibrated through a revised notion of labor that understood people as continual self-investment machines, or atomized sites of productive potential or diminishing returns within a free-market social context. In this sense, human capital theorists needed to come up with a new language and interpretive tools with which to erase the inherent contradiction of labor within industrial capitalist conditions of production where extractable value from the wage–labor process defined the worker–owner relation as well as state institutions that legally sanctioned them. Stated in Marx's language, human capital theorists needed to ask how the problem of alienation (one's productive forces set against oneself through the commodity form of wage labor) could be recast into a question of personal investment and entrepreneurial acumen—in effect shifting the question of class antagonism to the realm of individual responsibility and rational decision-making ability based on one's capacity to conduct his/her life effectively through sound economic thinking and behaviour in society. Indeed, moving the exploitative nature of work in the capitalist system of production into a framework of personal responsibility tied to market rationalities is one of the most important contributions human capital theorists have made to the neoliberal project. As Samuel Bowles and Herbert Gintis (1975) put it in their classic Marxist critique of human capital theory, the "degree of success [of human capital theory] is secured at a considerable price: 'labor' disappears as a fundamental explanatory category and is absorbed into a concept of capital in no way enriched to handle labor's special character. One gets the uneasy feeling that the operation was successful, but the patient vanished!" (74).

Second, and related to the first problematic, human capital theorists also needed to invent sophisticated analytical metrics in order to be able to measure how investments in human capital such as education had positive or negative effects on the economic value of individuals and populations in society. Human capital theory, in other words, required a new science of human economic value that could produce evidence to support the view that free-market economics and human ontology were indeed a Darwinian coupling. Through advanced statistical and mathematical analyses that focused on measuring inputs such as skill acquisition, education and training, dietary habits, ability to move to new labor markets, and mating choice, a truer picture of humans as income stream incubators could be offered and established as the foundation of social life and thus social policy formation. The filtering of people through human capital valuation metrics, put differently, made individuals and groups understandable as economic input actors on a stage where human nature seemed to have only evolved from an accountant's imagination.

One of the best examples illustrating how human capital theory attempts to make human ontology and free-market rationality into genetic partners is the Chicago School's value measurement technology cliometrics. Cliometrics is a statistical and mathematical model that was designed by human capital theorists to measure the ability of individuals and populations to respond and adjust to market cues embedded in capitalist social environments. Commenting on Robert Fogel's part in the revolution cliometrics caused within the broader field of economics, Claudia Goldin (1995) states, "The new economic history, or cliometrics, formalized economic history in a manner similar to the injection of mathematical models and statistics into the rest of economics" (193). As we will see in this chapter, humans living in social settings structured through capitalist relations, for human capital theorists, naturally develop an ethical orientation geared toward continual self-investment practices in areas such as skill acquisition, housing, education, mating choices, genetic endowment, health inputs (i.e., diet or access to good health care), and other types of human capital that help optimize a person's life-value in capitalist societies. Again the sleight of hand here embedded within human capital theory, and especially apparent in measurement tools such as cliometrics, is the substitution of internal crises that are a product of the social organization of labor under capitalist conditions with a problem of deficiency or impurity in the individual or population, namely, their value as investible units and productive assets in a globally competitive economic environment. Cliometrics was one such tool developed by neoliberal theorists and, as such, is a highly useful subject of analysis in that it was designed to perform precisely what was needed most: the translation of individuals and populations into measuraeable market beings. The transformation of all human life to fit the genome of *Homo economicus* also required the development of an educational dimension that has yet to be studied.

Neoliberal theorists in the United States have had a deep interest in education since the post-World War II period, and it is no surprise that the radical restructuring of the public education system has been one of the primary areas of their collective research.[5] Yet what is it exactly that makes education so important to the neoliberal project? One way to answer this question, and what my analysis in this chapter of an early human capital study attempts to do, is examine how neoliberal theorists construct a model of education out of their application of performance metrics to socioeconomic settings. Education, set within human capital frameworks such as Robert Fogel and Stanley Engerman's case study *TC*, is turned into a productive space where the individual and population are treated as investable and entrepreneurial sites inextricably connected to economic growth and market competition. Human capital models of education thus are subject to and even call for governing techniques that can optimize the quality of human capital within a nation's population, especially in times of economic and social crises such as our own.

In an effort to better understand how human capital theory undergirds the neoliberal restructuring of education currently underway in the United States, I take up two case studies in its developmental history in this chapter and the next. In this chapter, I revisit Fogel and Engerman's human capital study of slave plantations during the antebellum period in the United States, which, I argue, served as a sort of test run in which to unleash new economic "truths" into economic and social policy debates through the magic of a new breakthrough pseudo-scientific research method known as cliometrics. As a statistical and mathematical methodology, Fogel and Engerman posit in their infamous *TC* that cliometrics provides the most accurate picture of the productive value of African–American slaves in the antebellum economy from any previous study on the subject.[6] While Fogel and Engerman's study was deservedly met with considerable critique as it advanced an argument that erases the racial and gender violence endemic to the plantation system (Gutman 1975, Crowe 1976), their analysis is also distinctly representative of one of human capital theories' implicit traits: isolating and valorizing human life through measurement techniques such as cliometrics replace deep ethical and political questions by substituting in the figure of *Homo economicus* as the archetype of human moral action.[7] Ulrich Bröckling (2011) nicely captures how at the heart of human capital theory lies a Frankenstein-like hubris that desires all humans to behave like the creature *Homo economicus*:

> [The] theory of human capital grasps the human being as *Homo oeconomicus* and grasps him only to the extent that he behaves accordingly: if individuals constantly try to maximize their benefits, their actions can be guided by raising

> or lowering their costs and thus altering the calculation. As someone who constantly decides, *Homo oeconomicus* is also "someone who is eminently governable." (Foucault 2008, 270) [citation from Bröckling]
>
> If there is no behavior that cannot be described in terms of cost-benefit caluculations, then people have no other choice than to make choices in all their actions. The economic approach addresses them from the start as the entrepreneurial market subjects into which they need to be transformed and to transform themselves. (258)

Indeed, Fogel and Engerman argue that their human capital findings that fit Bröcklings above description do not answer questions about race relations in the United States. What human capital analysis does provide, according to them, is a framework that opens up a discussion on "questions [such as] as the nutritional adequacy of the slave diet, the profitability of investments in slaves, and the efficiency of slave labor" (Fogel and Engerman 1974, 9–10). As such, human capital studies such as Fogel and Engerman's also carry with them a proto-eugenic quality that has also dropped from sight in current debates surrounding educational reform, specifically reform approaches pushing assessment and standardization as a way to measure value-added life in schools. In the fundamental desire to maximize the entrepreneurial capacity of individuals and populations, human capital theory is ultimately interested in constructing social and institutional spaces where *Homo economicus* can flourish and where other forms of social life wither. Fogel and Engerman's *TC* demonstrates how not only such an environment can exist but more importantly how unfettered capitalist social forces unleashed on economically developing spaces provides the pedagogical cues for realizing the genetic potential in humans thought of as *Homo economicus*.

Michel Foucault was the first to point out the eugenic dimension of human capital theory that arose out of the Chicago School economics brand of neoliberalism (Foucault 2008). While Foucault only gestured toward a future where a politics based on human capital understandings of people would lead to the arrival of a biopolitics based on racial and individual genetic traits, he nevertheless clearly recognized this biopolitical tendency operating within human capital theories and governmental strategies organized around ideas of innate intelligence and genetic endowment. Foucault (2008) remarks, "as soon as society poses itself the problem of the improvement of its human capital in general, it is inevitable that the problem of the control, screening, and improvement of the human capital of individuals, as a function of unions and consequent reproduction, will become actual, or at any rate, called for. So, the political problem of the use of genetics arises in terms of the formation, growth, accumulation, and improvement of human capital. What might be called the racist effects of genetics is certainly something to be feared" (228). Such a prescient claim, I argue, precisely reflects the danger inherent in human capital theory where "all human abilities [are] either innate or acquired" and "every person is born with a particular set of genes which determines his innate ability" (Schultz 1981, 21). How this eugenic feature of human capital has evolved through new manifestations such as value-added metrics (the subject of the following chapter) will become clearer once its developmental roots are examined in works such as Fogel and Engerman's.

Along this same line, I argue that Fogel and Engerman's early human capital study rests on a biopolitical ground that is highly relevant to understanding contemporary educational reform strategies that have wholeheartedly adopted the human capital framework. Specifically, the human capital view that human beings retain varying degrees of innate (biological) qualities that give some a competitive advantage for increasing one's human capital while, for others, their genetic disposition exists as an impediment to human capital accumulation, is rooted in the understanding that individuals and populations are susceptible and even call for the creation of free-market forms of governmentality that increase the value of their life. A whole trend in human capital research in fact focuses on questions of "innate" dispositions of individuals as it relates to economic life, family, fertility, and marital choice—in essence, human capital theorists are deeply interested in the question of how individuals identify and invest prudently in the best available genetic stock so as to have a positive effect on their future income stream and, in turn, their life span. Perhaps the Chicago School's most well-known theorist and activist Milton Friedman (1995) puts the proto-eugenic feature of human capital theory most bluntly in his argument for school vouchers: "Innate intelligence undoubtedly plays a major role in determining the opportunities open to individuals. Yet it is by no means the only human quality that is important, as numerous examples demonstrate. Unfortunately, our current educational system does little to enable either low-IQ or high-IQ individuals to make the most of other qualities. Yet that is the way to offset the tendencies to stratification" (343). Here, the biology of humans and neoliberal strategies for restructuring education are married through radical privatization proposals such as Friedman's. Voucher and charter systems of schooling are clearly addressing the problem of regulating and managing life in society by introducing radical free-market governing policies to make the best of the innate qualities possessed in the population of the United States. In the case of Friedman's school voucher proposal, the radical free-market individual can only be unleashed by privatizing the public school system, which, in turn, can realize the productive value of the population while disposing of its less-valuable stock. This genealogical line must not be overlooked in the development of human capital theory within the neoliberal revolution over the past 30 years and is one I pick up again in the next chapter.

Fogel and Engerman's *TC*, as I hope to show in this chapter, is an indispensible guide for better understanding contemporary variations of human capital theory, in particular value-added metrics that comprise one of Race to the Top's most controversial features. There are two reasons I am situating my analysis of Fogel and Engerman's human capital study as the first case study of these two chapters. First, it is important to recognize the genealogical route taken by human capital theorists that has paved the way for reform approaches such as value-added measurement. By looking at early experiments in human capital theory development, I argue that we can better situate and understand value-added metrics as a knowledge system and diagnostic tool of educational performance. Second, I also argue that human capital theory has taken a distinct evolutionary jump in the biocapitalist era. In the neoliberal educational project that has optimization of the country's stock of human capital (STEM areas, for example) as one of its primary goals, a mirroring of other economic fields interested in producing valuable forms of biocapital is apparent. At one level, I am suggesting, there is little difference between genetic scientists' sorting and manipulation of DNA sequences into valuable and non-useful material ("junk DNA") and human capital strategies for obtaining the most value possible from educational populations while discarding non-vital biocapital. Ultimately, then, Fogel and Engerman's study of slave plantations and the type of market learning their study asserts is a particularly apt blueprint from which to interpret current human capital reform approaches in education that transform questions of social power relations and equity into a matter of free-market expansion and human capital accumulation.

Indeed, the focus on life, or its total economization, is a generative theme of human capital theory and spans the range of studies its researchers have developed over the past 60 years. For example, Theodore Schultz, one of the most important early pioneers of human capital theory, specialized in the area of economic and agricultural development in the developing world. Robert Fogel's research also remains along a trajectory focusing on questions of population health such as quality of nutrition and rates of hunger in advanced industrialized nations as well as industrializing countries. From this context, Schultz, Fogel, and Engerman render life in the field, village, school, and nation all as sites of investment and value extraction, or, perhaps more accurately, as potential zones of biocapital production. Value-added reform, the latest incarnation of human capital thinking, indeed has a dark family past that is not so easily swept away through statistical sanitation: on final analysis, I argue, life in the human capital universe is ultimately understood as an optimizable biological unit of production. In other words, the very goal of value-added measurement as a human capital diagnostic tool is to determine the worth of life in educational institutions and how to adjust for inhibitors not amenable to market understandings and valuations of human life. Excavating the origins of such an approach is even more important ever since human capital theory has become both the language and grammar in which educational policy and curriculum are increasingly required to speak. The best way to see where such a language was formed is to examine one of human capital theory's most important experiments.

Education and the Plantation: A Neoliberal Case Study in Market Learning

The choice of Robert Fogel and Stanley Engerman's *TC* as a place to begin reassessing the origins of human capital theories of education is not meant to drag up an old and for many a deeply painful controversy.[8] The preposterous claims made in their book remain just that: a delusional neoliberal study of African-American slave labor and the economic conditions of the antebellum South rendered through a pseudo-scientific methodology known as cliometrics.[9]

Rather, my goal in this chapter is to illuminate the genealogical path of human capital theory that seems to either have been forgotten or, perhaps more likely, its views seem less offensive to many in our "post-racial" society where questions of equity (most importantly in state and institutional policies) have been subsumed into free-market discourses and practices. Amnesia to such a past is an especially disturbing turn of events given the fact that human capital theory has become the leading framework that informs both educational policy debates and attendant techniques of measurement driving the current reform movement in the United States. Neoliberal quantitative techniques such as cliometrics are designed primarily as a way to measure and assess the economic value of human beings within specific social settings. They are also, however, rooted in a scientific truth regime that creates its own universe of data and research techniques that have as a tribunal the neoliberal fantasy of a free-market utopia. In the case of Fogel and Engerman's *TC*, the statistical and mathematical technique of cliometrics used in their study allowed them to reconfigure slaves in the antebellum period into subjects of human capital investment and accumulation as well as what I am calling *market learners*. Within the Chicago School's project to legitimize human capital research methods (which also included deadly experiments in nations such as Chile under the bloody rule of Pinochet), Fogel and Engerman's cliometric analysis of slave labor in the United States is far from being an anomaly within the neoliberal laboratory of the Chicago School (Valdez 2008, Klein 2008).

It is also not a coincidence that Fogel and Engerman chose the institution of slavery to launch their neoliberal revision of one of the darkest periods of the United States. If they could prove that free-market forces have always been a liberating and equalizing force in US society, even in times of great unfreedom such as a condition of legalized slavery, the road to social justice and racial equity could be rerouted through a convenient neoliberal detour. The choice of using the institution of slavery as a case study in which to showcase how human capital theories could transform even the most oppressive conditions for social life in the United States into a productive and pedagogical space thus had a disturbing ulterior motive. By removing the struggle for racial equity and economic justice from communities and individuals who have historically suffered under racist and capitalist systems of oppression, human capital theory sought to replace emancipatory cultural experiences and standpoints with sanitized versions of economic history that transform market forces into the source of greater freedom and justice over cultural knowledge and practices forged in spaces of great unfreedom and brutality.

There was also another callous reason Fogel and Engerman chose the slave plantation as the object of their study. For them, as well as other human capital theorists, slaves represented a pure form of human capital—an individual who was a property of another provided an important control variable for understanding the mechanisms of investment that slave owners used to decide how to increase the productive value of their property. In short, the institution of slavery offered Fogel and Engerman a model in which to test the views of human capital in what they saw as the most controlled conditions. As Gary

Becker coldly put it, a Chicago School economist whose work on human capital and education is still widely regarded, slavery "is the one example of an explicit market that trades and prices human capital stocks rather than simply the services yielded by these stocks. A major and insightful study [Fogel and Engerman's *Time on the Cross*] has recently appeared that interprets the market for slaves in the United States in terms of the theory of investment in human capital" (Becker 1994, 8). Comparable to the morally corrupt claims of eugenic science and its experiments in the first part of the twentieth century, human capital theory and the governmental strategies that engender its perspectives retain a decidedly biopolitical project. Life, seen through measurement devices such as cliometrics, becomes a "stock" to be managed and regulated through market strategies and pedagogies that teach individuals to conduct themselves in society as self-interested economic actors. Human capital theory is thus built around valuing and regulating forms of life within a predatory and ever-expanding economic framework that, as Foucault (2008) defined it, seeks to "exten[d] economic analysis into previously unexplored domain[s]" (219). As the primary extractive tool used to normalize the view that people are investible and market-driven subjects, Fogel and Engerman's human capital study exemplifies how measurement techniques reconfigure questions of equity and the role of education in society into a matter of market freedom.

There are two key features of human capital theory I identify in my analysis of Fogel and Engerman's *TC* that point to the biopolitical dimension of their project, or, the way educational life is a core target in their cliometric study. The first biopolitical aspect of *TC* I focus on is the pedagogical quality that cliometrics takes on as a measurement and sorting device used to determine specific types of human economic value in social settings. Cliometrics, in other words, is more than just a statistical and mathematical mega-machine; it is also a truth regime designed to support and advocate governing strategies for individuals and populations that promote *Homo economicus* as the ethical model of self-care. Second, what I am interested in is how human capital theory within Fogel and Engerman's study of slave plantations provides a strong case study in which to look at how neoliberal governmental strategies deal with the question of equity in sociohistoric settings. As such, Fogel and Engerman's *TC* offers insight into the fundamentalist belief in market responses as the most appropriate method in which to deal with social conditions where equity has not been achieved and in fact is never really a truly desirable outcome, at least not one that was not centered on a notion of market freedom. As I argue later, Fogel and Engerman's human capital study works on both of these co-articulating lines, and thus understanding each separately gives a clearer account of how *Homo sapiens* is transfigured into *Homo economicus* through the prism of cliometrics.

Cliometrics, or How to Find Value in People

As Herbert Gutman has pointed out in his devastating critique of *TC*, one of the overarching claims made by Fogel and Engerman is that the system of slavery was, contrary to all economic historical studies of the plantation system until then, actually a site of positive social learning for African–American slaves as well as slave owners.[10] Gutman (1975) notes, "So goes the main thrust of T/C. A mass of evidence—some of it old and some of it new, some of it quantitative and some of it literary—seeks to show how well the slaves learned from their owners about how to labor efficiently and productively (the 'Protestant work ethic') and about how to live in normal nineteenth century families (the mores of Victorian family life)" (6). The important question here is not the validity or accuracy of Fogel and Engerman's study. Indeed, Gutman and others have already convincingly dismantled the flimsy construct that supported the quantitative claims underlying *TC* years ago (Crowe 1976). What I am interested in this section rather is the question of how the methodology and framework that constituted such an erroneous and off-base interpretation of economic history are illustrative of human capital descendants and their analyses of educational problems today? Part of the answer to this question, I contend, has to do with human capital measurement tools such as cliometrics that understand individuals and populations through what Fogel and Engerman (1974) call a "series of rapid advances in economics, statistics, and applied mathematics, together with the availability of high-speed computers" that enabled them claim on a "scientific" basis that "the main features of the actual operation of the slave economy are now clear" (4).

As Charles Crowe (1976) remarked on the arrival of cliometrics two years after its publication, "*Time on the Cross* bristled with the heavy artillery of modern technology, including electronic scanners, high speed computers, intricate print-outs, the aerobatics of inferential analysis, the algebra of population genetics, and mathematical equations of Byzantine complexity" (592). In many ways, Fogel and Engerman's revisionist history of plantation economics was the inaugural launch of a massive "truth regime" machinery in which the cogs held their own self-fulfilling prophesies: that is, looked at in the right way and with the correct instruments Fogel and Engerman's study asserted, slavery was not only a profitable economic system but also a site in which economic value was intimately tied to a social pedagogy linked to the growth of capitalism in the Southern economy of the antebellum period. In returning to Fogel and Engerman's analysis of the social learning process that they argued was embedded in the condition of slavery, an important lesson in the development of human capital theory and practice can be derived: humans are, on final analysis, pure economic subjects and thus their ability to learn can only be determined if economic tools of analysis are used to scrutinize the potential and real entrepreneurial and investment habits of individuals and populations within capitalist social settings.

It was not lost on Fogel and Engerman that their revisionist economic history of the US slave plantation system would be considered controversial or even explicitly racist by many. In confronting such an interpretation of their work, Fogel and Engerman offered a rather bold preemptive explanation: their study of the economic conditions of the slave plantation in the Southern United States based on a new system of economic measurement, in fact, offered the most accurate and nonracist account of labor and life in the antebellum period.

In their own words, "we have attacked the traditional interpretation of the economics of slavery not in order to resurrect a defunct system, but in order to correct the perversion of the history of blacks—in order to strike down the view that black Americans were without culture, without achievement, and without development for their first two hundred and fifty years on American soil" (Fogel and Engerman 1974, 258). Also, in Fogel and Engerman's liberatory framing of cliometrics, an important discourse of human capital theory is apparent. Through the measurement tool of cliometrics, an equitable space is seemingly opened up that places African–American slaves on an equal (in fact, better in many cases according to their comparison between white industrial workers' wages and living conditions) productive plateau as other working groups of the period. Thus, for Fogel and Engerman, when human capital theory is applied through cliometric analysis, a positive reassessment of slave life is achievable and, in fact, offered as objective truth. Of particular importance to Fogel and Engerman is the social learning process that emerges in the shift from an aristocratic agrarian economy to more capitalist forms of labor instantiated through the industrialization of the south. Yet the question remains: How did Fogel and Engerman go about magically discovering a positive educational dimension to one of the darkest periods in US history?

Human capital metrics such as cliometrics, at the simplest level, operate on at least two important registers that need to be considered in order to fully understand Fogel and Engerman's interpretations of the type of positive learning that occurred in plantation life during the antebellum period. First, as a measurement and predictive tool, cliometrics fulfills a core goal of human capital theory, which is to translate human life into a series of economic investments that ultimately affect the earning profiles of people over their life cycle.[11] In the case of Fogel and Engerman's human capital study, cliometrics served the methodological function of being able to parse out the investment behaviors slaves and slave masters were learning through increased market incentives brought on through the industrialization of the south. By producing copious amounts of statistical and mathematical data that depicted the plantation system as an incentive-based social structure promoting positive individual gains as well as overall cultural advancement for the African–American slave population, Fogel and Engerman were attempting to introject the view that slaves on the plantation were incipient investment machines—individuals gradually morphing into *Homo economicus* as the feudal-like structures of the agrarian economy dissolved against capitalist progress. Second, human capital theory also carries a proscriptive or sovereign dimension in that market rationalities are presented as the natural logic by which governments should manage a population and how individuals should conduct themselves within society. Taken together, human capital theory contains a theory *and* practice of social life that delineate and translate almost all human decisions and behaviors into economic terms while maintaining that even the most intimate and basic acts such as marriage, diet, education, fertility, and genetic endowment are types of capital of which the value can be maximized by learning from the market.

A prime example of how Fogel and Engerman's human capital theory unites economic rationality with how individuals learn in their daily lives from the market can be drawn from Gary Becker's work on education and human capital. Becker (1994) argues in his book-length study on human capital and education that education includes not only formal education (schooling, college, or vocational training) but also on-the-job training and ability to move geographical locations for better work opportunities. Thus, Becker's human capital framing of education at its most general level can best be understood as social zones of learning that exist in capitalist societies and couple with the innate "ability" of individuals (measured through IQ testing and level of education attained by one's father, for instance) to have a direct positive effect on rates of return to one's income stream. This same concept of education, individuals learning from the types of demands the capitalist economic system puts on their lives, is present in *TC* and, as such, it is highly instructive to understand how Fogel and Engerman utilize this very broad notion of education as incentive responses in different social environments to help make their argument that plantations, viewed through the lens of cliometric analysis, were positive learning environments as well as productive economic ones.

Unlike earlier critiques of Fogel and Engerman's *TC*, I am arguing that the peculiar pedagogical dimension inlaid within their human capital analysis of slave plantations is operating on a distinctly biopolitical plane. In constructing a model of the individual and population that actually flourished and learned from the socioeconomic conditions under slavery, Fogel and Engerman are interested in how the habits and desires of subjects are produced. For them, and human capital theory in general, the biological destiny of individuals and populations is inherently connected to the capitalist organization of society in that one's quality of life depends on how well the habit formation of a subject is connected to learning from market-shaped social environments. In other words, it wasn't the sovereign power of the slave owner or the legal institution of slavery that primarily shaped the social life of slaves according to Fogel and Engerman, rather it was the growing market structures influencing slave owners and states in the south that created the most important areas where social learning for slaves occurred. Fogel and Engerman (1974) make this point in their educative representation of the exploitative system of work under slavery:

> Neglect of the fact that more than one out of every five adult slaves held preferred occupational positions, which involved not only more interesting and less arduous labor but also yielded substantially higher incomes, has encouraged still another oversight: that is, the failure to recognize the existence of a flexible and exceedingly effective incentive system that operated within the framework of slavery. The notion that slaveowners relied on the lash alone to promote discipline and efficiency is a highly misleading myth. In slave, as in free society, positive incentives, in the form of material rewards, were a powerful instrument of economic and social control. (40–41)

Fogel and Engerman's claim that slaves in fact responded to and socially advanced through market mechanisms under the institution of slavery (even more so than from violence) needs to be looked at carefully as it is a foundational tenet of human capital theory and one of the ultimate truth claims exposed through a cliometric analysis of slave life.

Connected to Fogel and Engerman's biopolitical model of market learning on the plantation is also a distinctly neoliberal notion of equity. In designing their free-market educational zone under slavery, Fogel and Engerman strategically situated their cliometric-based argument as one working on behalf of greater social and economic equity. In fact, they claim that their analysis of the economic and social conditions of slaves living under the plantation system offers a more egalitarian and nonracist view of African–Americans than most previous historical accounts up to that time. Fogel and Engerman moreover argued that their cliometric analysis corrected racist views being advanced by even groups and individuals who were against the system of slavery such as abolitionist writers and historians of slavery in the United States (Fogel and Engerman 1975). According to Fogel and Engerman (1974), then, "slave agriculture was not inefficient agriculture. Economics of large-scale operation, effective management, and intensive utilization of labor and capital made southern slave agriculture 35 percent more efficient than the northern system of family farming. The typical slave field hand was not lazy, inept, and unproductive. On average he was harder-working and more efficient than his white counterpart" (5). It is important to recognize that Fogel and Engerman's human capital analysis takes on the appearance of a discursive strategy of equity, which, as I argue in the next chapter, persists in current versions of human capital metrics. In other words, Fogel and Engerman's claim that their cliometric study challenges the "the racist premise that blacks were biologically inferior to whites" reproduced by both critics and defenders of slavery is where a particular meritocratic ethic emerges as an important aspect to their cliometric study. Specifically, equity in *TC* is translated as an individual's or a group's ability to increase the economic efficiency of their daily lives through skill acquisition and incentive-driven decisions compelled by external competitive labor forces affecting both large- and small-scale agricultural producers. The driving force behind increased social equity in the United States, even during a time of legalized slavery, therefore in Fogel and Engerman's cliometric universe was economic incentives that drove investment decision-making processes in individuals which occurred at the level of job status (manager or picker), learning and performing a new craft or skill valuable to the increasingly technological approach to agriculture (new farming and sorting equipment) which all had the goal of increasing one's economic standing and by extension their health (a longer life).

The influential role of pecuniary incentives (tangible monetary gains) in the plantation economic system, for Fogel and Engerman, was, not surprisingly, one of the biggest oversights made by previous studies of the slave economy in the United States. In fact, Fogel and Engerman, as I have pointed out above, considered slaves as pure examples of human capital, that is, the ownership and condition of their lives were entirely in the hands of the slave owner. Thus slaves' lives represented an economic unit in which to understand the merits of both human capital (how slaves recognized and benefited from market forces) and market forces that commanded a more equitable response from slave owners to their slaves as an investible and productive unit. From this view of slaves as pure forms of human capital, Fogel and Engerman (1974) assert that "far from being kept at the brink of starvation, slaves actually shared in the gains from economies of scale—so far as purely pecuniary income was concerned... This finding underscores the importance of the system of pecuniary incentives discussed in Chapter 4. It is not correct to say, as one leading historian did, that though pecuniary 'rewards were not much used,' force was the 'principal basis' for promoting the work of slaves. Pecuniary incentives were no more an incidental feature of slavery than force" (239).

Ensconced in Fogel and Engerman's argument for the decisive role financial incentives played in the creation of slave and slave owner subjectivities is an insidious notion of equity: the source of greater equity (at least in terms of pay and work roles) operating within Fogel and Engerman's cliometric analysis is a religious-like belief in the "invisible hand of the market" to be the best possible teacher and health consultant for both the oppressed and the oppressor. On the one hand, according to Fogel and Engerman, market forces inherent and exterior to the plantation economy compelled slaves to invest in certain skills and modes of behavior that lead to better and longer lived lives (better health options, less physically damaging work, etc.). On the other, the slave owner, despite his moral deficiency, was also compelled by pecuniary incentives to make sure his slaves were rewarded just enough to make the productive system of the plantation more profitable than competing small and large agricultural farms in the North that operated through the wage labor system or family-run enterprises.

In this sense, what appears in Fogel and Engerman's human capital study of slave plantations is this important biopolitical imposition: slave life organized, structured, and controlled through the free-market transformation of the southern economy during the antebellum period proved to be the most decisive factor in creating a more equitable arrangement in the social life of slaves. Moreover, the educative function of the market also had an overall positive effect on the entire slave population in *TC*. For instance, under a market-structured plantation system, slaves were able to learn how to be good and productive workers, what Gutman (1975) has pointed out in his critique of *TC* as an adoption of white southern aristocrats' "Protestant work ethic" by African–American slaves. Here, equity does not include original and powerful contributions of African–American culture to society as a whole, rather it means that modeling and aspiring to the codes of white society and the industrialization of work habits were the only path political life could take according to Fogel and Engerman's cliometric analysis. Here I am asserting that Fogel and Engerman's human capital study of slave life operates in a way that immunizes threatening or "impure" parts of the population that could not be easily folded into the triumphant narrative of market models of learning (Esposito 2008, Lewis 2009, Lewis and Kahn 2010).

What I am suggesting is that Fogel and Engerman's cliometric account of the plantation and the model of market pedagogy it engenders needs to also be seen as an attempt to immunize the history and legacies of resistance movements (i.e., slave rebellions, slave escape networks, or abolition reform) that were a constituent product of the south's agricultural economy built on slave labor. By emphasizing market forms of education, what Fogel and Engerman's analysis of equity in *TC* relies on stands in stark contrast to other prominent historical evaluations of the antebellum and Reconstruction period that represent an entirely different epistemological legacy and by extension collective learning model. W. E. B. Du Bois, for example, draws a uniformly opposite conclusion to Fogel and Engerman in his *Black Reconstruction* where he suggests that institutions and forms of social life maintained during slavery and the Reconstruction period should have been abolished if a serious effort toward building an equitable and democratic society were to be established after the civil war (Du Bois 1999). One of the chief reasons Du Bois reached such a conclusion in looking at the same historical period as Fogel and Engerman is because he viewed the legal and economic systems such as sharecropping, a caste wage system of pay where African–American and other workers of color were paid less than their white counterparts, and Jim Crow as a continuation of the violent and oppressive institution of slavery, not as a steady and progressive liberation of African–Americans through the south's greater embrace of the capitalist system. In short, Du Bois's analysis of slavery and Reconstruction reform efforts in the south suggests not social immunization, as does Fogel and Engerman's, but rather a spread of emancipatory views of social and community health based on the abolition of institutional and cultural systems rooted in extractive and oppressive relations with human life.

Hardt and Negri, more recently, have argued that Du Bois's body of work, which situates African–American culture as a wellspring for resistive modes of politics and social movements based in equity, provides an alternative entry point to sovereign forms of state power. For Hardt and Negri, Du Bois offers a strong example of forms of resistance that starkly contrasts with Fogel and Engerman's interpretation of social learning and provides an entirely different genealogical educative model based on counter knowledge and practice that have developed against hegemonic systems of oppression. Hardt and Negri (2009) argue that "slave resistance is a force of antimodernity not because it goes against the ideological values of freedom and equality—on the contrary, as Du Bois makes clear, slave rebellions are among the highest instances of those values in modernity—but because it challenges the hierarchical relationship at the core of modernity's power relation" (77).[12] For Du Bois, the counter ontological and epistemological standpoints generated in communities and individuals who live and know what he famously theorized as the subjective experience of twoness are grounded in resistance practices to modern power relations such as white supremacy and industrial capitalism that co-evolved within the governing structures of the US nation-state. This counter zone of fused knowledge and being, however, one that escaped and freed slaves, abolitionists, and figures such as John Brown and Nat Turner inhabited, points to another fundamental feature of the human capital framework; specifically, how technologies of control (such as schools) play a key role in attempting to neutralize potential threats to neoliberal understandings and governing rationalities of society. The normalization of *Homo economicus* as the standard educational subject through neoliberal governmentality strategies of schooling, in other words, carries with it a crisis management dimension in that potential areas of resistance in the population must be regulated in order to "succeed in covering the whole surface that lies between the organic and the biological, between body and population" (Foucault 2003, 253).

In Fogel and Engerman's human capital history of the slave plantation, we can see how market forms of learning play a pivotal role in their story of smoothly transitioning slaves into a more equitable and just social order ordained by the spirits of free-market economies. Human capital theories of education such as Fogel and Engerman's, I am suggesting, therefore also serve as a model of social and political containment in that self-investment for human capital accumulation does not tolerate the development of habits and desires not quantifiable through economic metrics of valuation. Here, human capital education is clearly operating at the level of population management and regulation. "Letting live" or investing in those groups and individuals who retain high human capital potential while "letting die" or disinvesting in other communities that lack a promissory future as a vital resource for the globally competitive economy and are threatening to neoliberal power relations is one of the most important underlying biopolitical features of human capital education and lessons we can learn from Fogel and Engerman's study.

Population Health and Equity

In Fogel and Engerman's cliometric analysis of slave life, when looked at from the vantage point of immunization or the containment of resistant forms of life, a second biopolitical register of human capital theory is made apparent. If, on the one hand, cliometrics was a tool that allowed Fogel and Engerman to measure and assess individual slaves' economic value within capitalist structured social environments, population health measurement, on the other, is human capital theory's other pole used for determining how well groups adopt the behaviors and desires of market-governed society. Such a focus on population health (i.e., a group's economic potential as productive energy) is one of the pillars of human capital theory, and it is important to see how this analytic category operates in not only Fogel and Engerman's study but also within the entire cartography of human capital studies. Fogel's work in the field of economic history after *TC*, for instance, focuses on issues of health, disease, and hunger in the Western world from a human capital perspective (Fogel 2004, Fogel et al. 2011). However, it was another Chicago School theorist whose work in the area of population studies from a human capital standpoint had the most influence on *TC's* analysis of plantation life.

Theodor Schultz's analyses of agricultural, health, and educational policies and practices in the developing world (as well as advanced industrialized nations) had a tremendous impact on the evolution of human capital theory and specifically how it makes sense of types of investments and governmental strategies for optimizing the quality of human capital in a population. As Schultz (1981) himself put it, "my approach to population quality is to treat quality as a scarce resource, which implies that it has an economic value and that its acquisition entails a cost" (12). It also includes the following:

> Treat[ing] everyone's state of health as a stock, i.e., as health capital, and its contribution as health services. Part of the quality of the initial stock is *inherited* [emphasis mine] and part is acquired. The stock depreciates over time and at an increasing rate later in life. Gross investment in human capital entails acquisition and maintenance costs, including child care, nutrition, clothing, housing, medical services, and care of oneself. The service that health capital renders consists of "healthy time" or "sickness-free time" which contributes to work, consumption, and leisure activities. (13)

Such an understanding of population health and quality was a primary analytic focus of Fogel and Engerman's cliometric study of plantation economics. For example, Fogel and Engerman's interpretation of slave populations in the south assumed Schultz's understanding of human populations as a "human health stock" that retained latent potential as an economic value if a certain amount of cost was invested into it, which, in turn, could elevate the "quality" quotient of the population. In framing types of quality attributable to certain populations within society, human capital theory categorically excludes alternative sources of quality by immunizing value systems that are inimical to market growth. In other words, what is measured in human capital metrics interested in population quality are not cultural practices and knowledges that questioned or challenged the institution of slavery such as prominent figures of the abolitionist movement such as William Lloyd Garrison or Fredrick Douglass as I pointed out above. Rather population quality, as a measure of community and individual health, has more to do with excluding variables not commensurate with market understandings of the potential vital energies of populations that Schultz's work presents as normalized in developed and "developing" countries once the proper institutional and legal means are established.

One of the primary ways Schultz's and by extension Fogel and Engerman's population quality analysis attempts to eradicate alternative social and cultural value systems is through an imposition of a type of artificial scarcity in calculating human valuation within population groups. That is, within the human capital framework, not all life is understood as equal (in both a genetic and an economic sense) and thus investment in forms of human capital such as dietary decisions, medical treatment options, choice of a mating partner, or education is recoded as scarce goods that quality life is measured by and in fact requires in order to increase its market value. Within Fogel and Engerman's framing of plantation life as a positive learning environment, the construct of artificial scarcity in terms of population quality is foundational and symptomatic of human capital theory writ large.

It is not a surprise then that the question of population quality existed as an important hinge point for Fogel and Engerman's analysis of slave life and work in *TC*. Their chapter "The Anatomy of Exploitation," perhaps the most important section to *TC*'s argument, relies on population health indicators to build their case for a positive account of the slave population in the antebellum period. In each of the subheadings of the chapter, "food, shelter, and clothing"; "medical care"; "the family"; and finally, "punishment, rewards, and expropriation," Fogel and Engerman reversed almost all previous historical interpretations of the condition of slave life in the south. By reconfiguring each of these quality of life indicators in their analysis of the population health of slaves, Fogel and Engerman presented an entirely new picture that showed slaves as a whole experiencing uplift in each of these areas under the increasingly efficient market-based plantation labor system. For example, Fogel and Engerman's cliometric analysis of the slave population produced a number narrative that ostensibly dispelled the "myth" that the dietary conditions of slaves were unhealthy. Fogel and Engerman (1974) argue: "Figure 33 ['created with data obtained from the manuscripts schedules of the 1860 census'] shows that the average daily diet of slaves was quite substantial. The energy value of their diet exceeded that of free men in 1879 by more than 10 percent." Fogel and Engerman go on to make the even more grandiose claim that "slave diet was not only adequate, it actually exceeded modern recommended daily levels of the chief nutrients" (113–115). Here, Fogel and Engerman are attempting to execute a crucial twist in their human capital analysis pertaining to the quality of the slave population in the South. In particular, that slaves, taken as a population, actually measured relatively high in terms of their dietary health. However, the more important and equally invalid lesson they were trying to draw was that the dietary health of slaves was a direct result of investments being made by slaveholders compelled by market forces and scarcity to increase the quality of the population under their control.

Again it is important to remember that my analysis of Fogel and Engerman's application of human capital theory to slave plantation settings is not necessarily about the validity of their data. The debate surrounding many of the inflammatory claims made in *TC* has long passed. Yet what has not seemed to pass is the power of economic metrics of human performance and in particular, how they have become one of the primary ways educational achievement and equity has come to be defined in the neoliberal era. What is still useful therefore about Fogel and Engerman's study is that it represents a blueprint approach of human capital models of social learning that needs to be interrogated, especially in contexts where defining equity and appropriate types of market governmentality is a key strategic area of discourse and practice driving the neoliberal educational restructuring project in this country. As I have argued in this section, one of the more important lessons to be drawn from *TC* is how human capital metrics participate in transforming something as unjust and violent as the institution of slavery into a positive learning environment indebted to laissez-faire growth. Population and individual health in the world *TC* depicts through cliometrics tries not only to completely integrate human life into the political economy of industrial capitalism, but also more accurately radicalizes it into a "biopolitical economy" (Bröckling 2011). Here, the population is subjected to and governed by a "generalized principle of utility maximization" and where "the individuals' approach to their own health as the consequence of decisions regarding investment and disinvestment" (Bröckling 2011, 261). It is a human capital constructed social environment that creates the ethical and political imperative where "blaming the victim here rules: whoever is sick has not adequately looked after his health; whoever falls victim to an accident or crime ought to have better seen to his own security" (261).

There are two primary lessons I want to emphasize from the biopolitical interpretation of *TC* I have offered and specifically its utilization of a human capital metric. The first is how the individual under the condition of slavery is ontologically cast as a learning subject who is most responsive to the entrepreneurial and incentivized economic structures of the developing capitalist relations of the antebellum south. The second biopolitical feature I am suggesting *TC* exhibits that is crucial to understanding current human capital metrics of education is how the quality of the slave population in the south is presented as achieving increased social and economic equity through an expansion of human capital opportunities provided by the capitalist system to a less developed economic *and* cultural system. But what is perhaps the most important aspect of Fogel and Engerman's study I would like readers to take away in this first case study is how human capital theory provides a framework for understanding and interpreting social life under slavery as a positive learning experience that could have only come about through the expansion of market governmentalities over individuals and populations that opened up unrestricted market access to more and more aspects of human behavior and habits. Herein lies the biopolitical bent to Fogel and Engerman's work and human capital theory in general: more than a purely ideological argument, *TC* presents a picture of social life under a system of extreme social control and regulation as an overall educative and fundamentally natural and equalizing condition. The habits and behaviors that slave masters, individual slaves, and the slave population embodied according to Fogel and Engerman's story are ones generated only through the complete subjectification of an ethic of self learned through the accumulation of greater "pecuniary income, diet, health, skill acquisition, and other aspects of the material conditions of life." It is this biopolitical framework, an understanding of life as a site of economic and social intervention, that descendants of Fogel and Engerman's study must be viewed today. In other words, how value in educational settings in our contemporary moment is measured cannot be thought of outside of the neoliberal project that Fogel and Engerman and other human capital theorists have helped develop. Instead, the question of how to value life in educational settings could not have been asked in the first place if the pieces had not already been set into place by studies such as Fogel and Engerman's. The next chapter will examine one of Fogel and Engerman's most successful genealogical descendants that has become both the most controversial and popular valuation metric of the post-NCLB era.

CHAPTER 2

Schooling for Value-Added Life: The Making of Educational Biocapital

Is not life more than meat, and the body more than raiment? And men ask this to-day all the more eagerly because of the sinister signs in recent educational movements. The tendency is here, born of slavery and quickened to renewed life by the crazy imperialism of the day, to regard human beings as among the material resources of a land to be trained with an eye single to future dividends.

—W. E. B. Du Bois (1903)

During the summer of 2010, the *Los Angeles Times* published an article that unleashed a firestorm of debate. According to the story's authors, their intent was to "estimate the effectiveness of L.A. teachers" across the Los Angeles Unified School District (LAUSD) using a new accountability tool that measured teacher effectiveness in relation to their students' academic performance. What made the article such an incendiary event was not only the *Times'* brazen use of seven years of test scores in math and English suspiciously obtained from LAUSD, but also the method chosen to analyze and depict the educational lives of students and teachers in the second largest school district in the country. Taking center stage, the *Los Angeles Time's* article cast the value-added method in the hero's role in one of its most important public test cases for the latest tool in the educational accountability movement that began in earnest with *A Nation at Risk* in the Reagan era, to George W. Bush's bipartisan policy No Child Left Behind and now Obama's Race to the Top which currently defines the educational horizon in the United States. The authors of the *Los Angeles Times* article introduced the main character, value-added metrics, by informing readers that "the Times used a statistical approach known as value-added analysis, which rates teachers based on their students' progress on standardized tests from year to year. Each student's performance is compared with his or her own past years, which largely controls for outside influences often blamed for academic failure: poverty, prior learning, and other factors" (Felch, Song, and Smith 2010, 1).[1]

Further into the story, the reader finds out what makes this new measurement tool so unique: "Value-added analysis offers a rigorous approach. In essence, a student's past performance on tests is used to project his or her future results. The difference between the prediction and the student's actual performance after a year is the 'value' that the teacher added or subtracted" (ibid., 4). Indeed, the *Los Angeles Times* article seems to deliver precisely the type of salvationary science that transforms the intractable problem of providing equal and quality education to one of the nation's largest urban school districts beset with longitudinal problems such as funding neglect, high teacher turnover rates, and the continuation of segregated schooling patterns into a matter of teacher performance.[2]

Yet there is also another important aspect to the *Los Angeles Times* showcase story on value-added metrics I want to draw attention to in this chapter. It signals the arrival of a fringe and controversial measurement device originally created in a statistics department at the University of Tennessee in the 1980s that has subsequently become one of the nation's leading diagnostic tools used for assessing and measuring the quality of life in schools. As the *Los Angeles Times* article notes, the value-added approach has been fully embraced by the Obama administration and, in particular, Arne Duncan's Race to the Top program, as well as by many district leaders who are implementing the assessment model as a way to drastically reform "underperforming" schools. For example, in the Washington DC school district formerly headed by the lightening rod Chancellor Michelle Rhee who now runs the education reform group StudentsFirst in California, value-added metrics were adopted and deployed to dismiss over 240 teachers (and counting) in the large urban school district.[3] The DC district webpage for performance and assessment extols their value-added driven IMPACT system, otherwise known as the DCPS Effectiveness Assessment System for School-based Personnel, that "helps us estimate the teacher's impact on student learning as opposed to the impact of other factors, such as students' prior skill level, the resources they have at home, or any learning disabilities they may have" (District of Columbia Public Schools 2011).[4]

However, another troubling question remains that the *Los Angeles Times* article elides: from where did this seemingly magical assessment model originate? Curiously, it is not until the fourth page of the feature story in the *Times* that the reader sees any mention of the origin of such a groundbreaking method of educational assessment. Here, the vague starting point for the value-added approach is said to have been "pioneered by economists in the 1970s" and "has only recently gained traction in education" (Felch, Song, and Smith 2010, 4). In fact, the story of value-added assessment, at least in the field of educational assessment, has a very discernable beginning that can be traced back to a familiar neoliberal mode of economic analysis.[5]

The value-added revolution in education began in the 1970s in the work of statistician and math professor William L. Sanders. Holder of degrees in biostatistics and genetics, Sanders began his research in the field of agricultural science where he applied agricultural growth models to educational assessment science. Under then Republican Governor Lamar Alexander, Sanders' early research on value-added metrics was part of a broader political project by state Republicans to implement a merit pay system for teachers in the state of Tennessee.[6] As Sanders himself recalls in a 1999 interview with the free-market think-tank the Heartland Institute, "I got involved when our legislators were told that you could not use student achievement data to measure teacher effectiveness. I said that you could, and got access to Knox County Student Achievement Test data to prove it" (Clowes 1999). In work that followed at the University of Tennessee, Williams and his new center for statistical research developed a "statistical mixed model theory and methodology" that would be the basis for what he called "value-added assessment."[7] Sanders and his colleagues went on to build the Tennessee Value-Added Assessment System (TVAAS) which was designed to "provide quantitative, reliable robust measures of schooling influences on the rate of academic progress of student populations" (Sanders 2007, 1). The TVAAS measurement technique is built on sets of sophisticated data analyses that, when run through complex software programs written by Sanders and his research team, produce what he has called a "river of diagnostic information to show teachers where their relative strengths and weaknesses are" (Clowes 1999). One of the most important factors leading to TVAAS's widespread adoption in the United States was the Department of Education's approval of Sanders' growth model as a method for "determining whether schools were making Adequate Yearly Progress under NCLB" (Darling-Hammond et al. 2012, 5). Since its incorporation in 1993 into the Tennessee educational accountability system and federal adoption through the NCLB policy, the TVAAS model (and similar ones) has spread to districts across the country including the states of Texas, North Carolina, South Carolina, Washington DC, Illinois, Wisconsin, New York, Pennsylvania, Utah, Hawaii, Ohio, Delaware, Florida, Arizona, and California, while Williams himself has been called before Congress a number of times as an expert to speak on the merits of value-added assessment.

From its early history as a little known assessment tool in the field of education, Sanders' value-added model has been lauded as the latest panacea in the growing high-stakes accountability movement in education. In particular, the value-added model is viewed by many as a remedy for No Child Left Behind's fatally flawed testing methods that rely on year-end test scores for measuring adequate yearly progress (AYP) in schools. Value-added proponents argue that their assessment metric is superior because year-end test scores lack the necessary data to determine the impact teachers have on students' individual learning—what should be the principle unit of measurement according to Sanders and others. Secretary of Education Arne Duncan, for example, has praised the transparency of value-added measurement asserting that "public disclosure of the value-added results would allow school systems to identify teachers who are doing the right things" (Felch and Song 2010, 1). The *Los Angeles Times* story is indeed correct in pointing out how "spurred by the administration, school districts around the country have moved to adopt 'value added' measures," which is also a requirement written into the Race to the Top legislation in order for states to qualify for federal funding. In fact, states such as Delaware and Tennessee (the birth place of value-added assessment in education), winners of

Phase I Race to the Top funding, utilized value-added approaches in their state strategic plan in part because value-added metrics are given priority in federal monies distributed by Race to the Top. To date, Race to the Top awardees reflect the strong preference for states that have in place a value-added assessment system as well as legislation for expanding the charter school system (Shear and Anderson 2009). Duncan and the Department of Education have also taken a carrot-and-stick approach in cases such as Washington DC where phase I funds of Race to the Top were denied to the district because the DC teacher union "opposed the new IMPACT teacher evaluation system." The DC school district was, however, awarded funding in Phase II of Race to the Top after the union accepted a new contract that weighed value-added scores in their performance rating (Turque and Anderson 2010). The meteoric ascension of the value-added model from a statistics department at the University of Tennessee to become the spearhead of the Department of Education's marquee reform policy is astounding however not suprising. Yet what is perhaps more incredible is how the value-added model uncannily resembles its human capital predecessor cliometrics, the subject of the previous chapter.

Indeed, the development of value-added metrics in the field of educational assessment is one that mirrors Fogel and Engerman's cliometric study I examined in chapter 1 which sought to derive and measure a free-market basis for social life under the condition of slavery. What is shared between value-added metrics and its antecedent cliometrics is the distinct human capital anatomy they both assume. A good way to see such an anatomical likeness in the value-added model is to look at the constituent actors involved in the power relation that value-added metrics relies on. The problem value-added metrics pose of how to measure the 'value' a teacher adds or subtracts from his or her students, in other words, comprises a power relation similar to that of Engerman and Fogel's cliometric study where slaves, slave owners, and the growing capitalist economy all coalesced into a natural market learning relationship. Fogel and Engerman's *TC* had *Homo economicus* emerging from a plantation system that was transformed by capitalist progress as a triumphant and market-educated soul. Value-added metrics work with a similar triad in the assemblage teacher/student/market forces. Perhaps the best way to understand the assemblage of teacher/student/market forces in the value-added model and how it mirrors Fogel and Engerman's cliometrics is to become familiarized with one of its closest cousins.

From Financial Capital to the Classroom

Value-added assessment techniques, as the *Los Angeles Times* article noted, also have a parallel history in the business and finance world known as economic value added (EVA) which is important to understand in considering the historical development of human capital metrics. EVA is a trademarked financial method developed by the worldwide management and consulting firm Stern Stewart & Company.[8] According to the company, EVA has become the most successful performance metric used by companies and their consultants across the globe for "improved economic performance and a higher enterprise value." Stern Stewart & Company's formula for calculating EVA consists of the "net operating profit after taxes and after the cost of capital. Capital here includes cash, inventory, and receivables (working capital), plus equipment, computers, real estate. The cost of capital is the rate of return required by the shareholders and lenders to finance the operations of the business. When revenue exceeds the cost of doing business and the cost of capital, the firm creates wealth for the shareholders" (Durant 1999, 2). If EVA is translated into its educational corollary, it is hard to ignore the genealogical resemblance to educational variations of value-added metrics.

For instance, by plugging in the actors that constitute educational value-added into the EVA formula, it quickly becomes evident that the most important variable from a shareholder or financial investment point of view is the teacher. The reason the teacher holds such an important position in this equation is because it is the role in production that retains an equivalent value to capital in the above definition of EVA. That is, the role of the teacher is the most important productive input in relation to the potential value that students (cost of capital) hold as investable units. It is the teacher, in other words, that is (1) a real cost in terms of salary, training, and benefits and (2) it is his or her labor that organizes working capital (curriculum, educational technology, classroom organization, grading and assessment, etc.) in the classroom provided by district, state and federal investments (and increasingly directly from the private sector). In this sense, just as in Stern Stewart & Company's EVA formula, the teacher (capital) is viewed as having the greatest influence on the outcome of the cost of capital (students) and in particular, the amount of value they can add to the positive side of the equation. Such a realization from a value-added perspective, of course, situates teachers as the primary target of control by investors and shareholders who want to see the potential value added of students maximized. The teacher input within value-added models, stated differently, is the crucial control variable in such a line of production and investment because his or her labor can be controlled most for student optimization and thus the human capital stock within schools.

In situating the value-added approach within the framework of corporate management culture, an important trend is brought to light. Value-added metrics are tools designed to help maximize returns on capital costs and to ensure a favorable profit margin for investors. The shared neoliberal mission of corporate interests and Race to the Tope educational reform where "out educating and out competing" the rest of the world is top priority demonstrates how educational life in the hyper-competitive global capitalist economy has clearly become another extractable field of value and strategic site for certain types of economic investment and intervention. Put differently, educational value-added metrics and corporate applications of EVA are symmetrical in that they both are used as a managerial technique of control that guides "financial institutions and natural resource firms, including oil and gas, mining and forest products" and for extracting "useful" educational life from the country's population (Stern Stuart & Company 2011). Whereas the extractive industries hunger for new forms of life to capitalize on such as raw resources like tar sands, natural gas, or the biodiversity of the Brazilian rain forest, value-added metrics in education seek to mine and maximize biocapital from educational populations.

The common link between value-added metrics in the finance and corporate world and that of neoliberal educational reform strategies is the shared human capital framework. Both value-added models normalize the total economization of life through discourses and practices such as those associated with EVA and similar corporate management technologies of control that understand employees, natural resources, students, and teachers all as forms of biocapital to be managed, regulated, and optimized for greater value. Value-added metrics thus transform educational settings into zones where value is extracted and optimized, rendering life in schools as a type of vitality to be invested in and shaped by market forces. Educational life subjected to value-added metrics, in other words, is objectified into something akin to a natural resource where educational vitality (the creative and productive energies of students and teachers) is understood as a minable and controllable resource to be efficiently extracted while non-useful material from the process of value maximization is discarded. Within this process of vitality extraction, value-added metrics also performs another important function in the educational biocapital production process. In a globally competitive education market certain forms of educational biocapital are targeted over others in the production of potential human capital surplus value. STEM areas of education, as I examine in Chapter 3, are a highly weighted form of human capital education valued more by economic industries which need individuals invested in immaterial forms of labor that are required in the laboratory, software design start-ups, or other high tech areas of the economy. Other forms of educational biocapital, on the other hand, become externalities, or disposable biomass along the production line of human capital optimization that schools in the age of biocapitalism have only helped accelerate.

Just as Fogel and Engerman translated the plantation system of the antebellum period into an equitable economic and social setting through cliometrics, the value-added framework similarly attempts to normalize a free-market ethical framework as the guiding moral force regulating and managing the parameters in which educational life should take place. Cliometrics allowed, in other words, Fogel and Engerman to argue that slaves benefited directly from the coalescing of social life under the institution of slavery and the valorization process embedded in the capitalist mode of production. For example, recall that Fogel and Engerman's neoliberal revision of economic history generated through the machinery of cliometrics needed to claim that all previous historical economic accounts of plantation life and work were inaccurate or "counterfactual." But their story was not just any revision of history; it was a complete overhaul of the historical conditions of slave life that sought to establish the basis to all human life (even in one of its most oppressed conditions) as *Homo economicus*, a human being who only learned and advanced in society by embodying free-market habits and behaviors—someone who conducted himself/herself in line with a competitive investment rationality increasingly informed by a social system of production where "his whole life, and that therefore all his disposable time is by nature and by right labour time, to be devoted to the self-valorization of capital" (Marx 1977, 375).

The value-added model, which also relies on a "complex and sophisticated" measurement technology, retains a similar need to cleanse historical and cultural "variables" such as race, gender, sexuality, and class inequities from educational life by turning them into external factors to the self-valorization process embedded in neoliberal educational strategies which understands individuals as a series of human capital investments. Its designers in fact claim, just as Fogel and Engerman did with their "amazing" new sorting technology cliometrics, that one of the greatest virtues of the value-added framework is its ability to isolate variables such as family income, race, gender, home culture of students, and other externalizable factors from the educational zone of human capital accumulation. However, both cliometrics and value-added metrics must be viewed as more than simply an ideological attempt to hijack definitions of how learning in sociopolitical contexts take place. Instead, both Fogel and Engerman's and the value-added framework, as I have pointed out, are biopolitical projects. In other words, they seek to *produce* subjects who embody laissez-faire habits and desires by controlling and regulating educational populations and individuals through governmental strategies that attempt to compel people and groups to embrace an ethic of self-care that views things such as competitive enrollment in high-achieving schools, disinvestments in failing urban schools, debt servitude to financial giants such as Sallie Mae, and merit pay for teachers as simply choices within a calculable economic field of decision making. It is important to see how such a biopolitical dimension operates in value-added approaches to educational life and how it is also part of the larger design envisioned by one of its primary engineers.

Decontaminating the Learning Experience: Controlling for Educational Value

In Sanders' description of how value-added metrics offer a purer picture of educational life (in particular, teacher efficiency in the valorization process of human capital accumulation) than class achievement test scores, he has pointed to the types of variables his metric systematically purges. These include "parental influences, genetic endowment, other school influences, and availability of materials" which, for Sanders, represent clouding influences that do not tell the true story behind student learning rates (Sanders and Horn 1994, 304). Sanders, not surprisingly, critiques evaluation models that fail to isolate and remove such factors from their analysis: "Any system that will fairly and reliably assess the influence of teachers on student learning must partition teacher effects from these and other factors. However, it is a hopeless impossibility for any school to have all the data for each child in appropriate form to filter all of these confounding influences via traditional statistical analysis" (304). Sander's value-added system, in contrast, claims that "these influences can be filtered without having to have direct measures of all the concomitant variables. By focusing on measures of academic gain, each student serves as his or her own 'control'—or, in other words, each child can be thought of as a 'blocking factor' that enables the estimation of school system, school, and teacher effects on the academic gain with the need for few, if any, of the exogenous variables" (305).

Sanders is not alone in the view that value-added metrics provide a kind of Rosetta Stone that staunch advocates of the metric-driven accountability movement in education have long been searching for. In his address to the Aspen Institute, a nonpartisan think-tank that has as its mission to "promote nonpartisan inquiry and an appreciation for timeless values," Ted Hershberg, professor of public policy and history at the University of Pennsylvania, echoes the increasingly popular call for value-added led educational reform. In his talk titled "Value-added Assessment and Systemic Reform: A Response to America's Human Capital Development Challenge," Hershberg praised the work of Sanders and others that have succeeded where other assessment models based on year-end test scores have failed: "If we are predicting student growth—progress made over the year—reports by education researchers Kain, Hanushek, Sanders and others demonstrate that *good instruction is 15–20 times more powerful than family background and income, race, gender, and other explanatory variables*" (Hershberg 2005, 5). But Hershberg (2005) also stresses that "the real prize in having value-added assessment widely adopted . . . is that it will change the organization and governance of K12 schools" where we will "have in place the means to bring all American students to internationally competitive standards and help meet the nation's human capital development challenge" (10).

The Bill and Malinda Gates Foundation has also been one of the most active players in designing, testing, and implementing value-added metrics in the classroom. David Labaree, for instance, notes that the Bill and Malinda Gates Foundation "has plunged $355 million into the effort to measure teacher effectiveness. Grounded in the value-added approach, this effort is using analysis of videos of teaching in individual classrooms to establish which teacher behaviors are most strongly associated with the highest value-added scores for students" (Labaree 2011, 10).

Given such an enthusiastic reception to the value-added model from powerful corporate groups and educational policy makers, it is not a surprise that research on value-added approaches has sprung into a cottage industry. Take, for example, the University of Wisconsin-Madison's Value-added Research Center (VARC) directed by Robert Meyer which is heavily involved in supporting value-added studies and implementation in educational settings across the nation. Meyers and his colleagues at the VARC have also sounded the alarm that traditional educational evaluation based on class testing is an inferior and inaccurate measure of educational outcomes.[9] Sharing Sanders' views on the methodological merits of value-added analysis, Meyer (1997) argues that

> this [average test scores] indicator suffers from four major deficiencies: it fails to localize school performance to the classroom or grade level; it aggregates

> information on school performance that tends to be grossly out of date; it is contaminated by student mobility; and it fails to distinguish the distinct value-added contribution of schools to growth in student achievement from the contribution of student, family, and community factors. As a result, the average test score is a weak, if not counterproductive, instrument of public accountability. The value-added indicator is the conceptually appropriate indicator for measuring school performance. (298)[10]

Here we can see that common to most value-added models is the desire to wipe free any "contaminants" such as race, family situation, or gender that fall outside of the control of schools and teachers; such variables are unimportant and, in fact, are considered misleading to student (and teacher) learning according to value-added logic. Again, recalling Fogel and Engerman's cliometric analysis that sought to determine the economic value of plantations and emphasized the beneficial learning experiences of free-market expansion by wiping clean the dehumanizing and violent truth of the slavery, proponents of value-added metrics similarly point to methodological hygiene as a design strength. Such a methodological purity claim made by value-added proponents should not be overlooked. If it can be asserted that science (value-added metrics) is able to neutralize and expel matters of race, gender, and class from the learning context of schools, variables that muddy precise assessments of the value of life in the classroom, then what does this mean for alternative and potentially resistive knowledges and standpoints to existing socioeconomic structures and cultural values that exist in classrooms and communities? Consider, for example, how the value-added approach for measuring students' learning jettisons pedagogical approaches such as the "funds of knowledge" framework. "Funds of knowledge," one of the bases for what Gloria Ladson-Billings has termed "culturally relevant pedagogy" in her work, would be viewed as a "contaminant" to the measuring of teaching and learning in the value-added framework. A value-added approach instead compels teachers to respond and deliver human capital inputs that enrich students with the appropriate type of value needed for the competitive economic milieu of global capitalism (Ladson-Billings 2009, Delgado Bernal 2002, Gonzales et al. 1995). Here, the transformative potential of students' lives as a positive contributor to learning and teaching is systematically barred from entering into the learning experience in schools and, in fact, is considered external to the valorization process of human capital production through education. What remains instead are market pressures designed to invest in "useful" human capital while disposing of cultural wealth and the transformative potential of marginalized students and communities from the educational process altogether.

In this sense, value-added proponents' desire to sterilize the classroom from cultural and historical experience I am suggesting should also be understood as a technology of control over populations. If part of the merit of value-added metrics as a measurement device is derived from its ability to isolate and externalize student attributes such as race and gender out of the valorization process of human capital education, another is the implicit goal of shaping school populations by creating high- and low-yield schools of human capital stock. One of the best examples that demonstrates how the adoption of value-added reform policy can be viewed as an effort to regulate educational stocks of human capital is the Washington DC school district.

The DC school district has implemented value-added metrics on two levels: individual measures of value-added (IVA) and school value-added (SVA) which are both integral to the district's IMPACT system built on teacher performance assessment (District of Columbia Public Schools 2011). Such a value-added rating system for schools, however, has a profoundly negative effect on students and communities despite Michelle Rhee and others' claim that school ratings through value-added scores empower parents to choose better performing schools over "weaker" ones. Since the institutionalization of Rhee's reforms that have systematically shut down underperforming schools and dismissed hundreds of teachers from the DC district, one outcome is clear. Segregation has increased in an already highly inequitable school system under the name of school choice and value-added scoring. Describing the manner in which Rhee's IMPACT system is reinscribing historic forms of inequity under the guise of school choice and rigorous new performance measurement techniques, a DC parent shares the stark reality her student and many others in the city face:

> There aren't enough slots in the best neighborhood and charter schools. So even for those of us lucky ones with cars and school-data spreadsheets, our options are mediocre at best. In the meantime, the neighborhood schools are dying. After Ms. Rhee closed our first neighborhood school, the students were assigned to an elementary school connected to a homeless shelter. Then that closed, and I watched the children get shuffled again... These proposals [replacing a closed school with a new magnet], like much of reform in Washington, are aimed at some speculative future demographic, while doing nothing for the children already here. In the meantime, enrollment, and the best teachers, continue to go to the whitest, wealthiest communities. (Hopkinson 2011)

The effects of value-added educational restructuring are not just a problem for communities in DC. In New York City, high-achieving urban schools are facing a competitive surge in terms of student enrollment since the New York City School District has adopted an aggressive value-added assessment model in the city's new school reform plan. Under the Bloomberg administration, school choice competition coupled with value-added scoring has led to a social Darwinistic selection process that is a result of schools needing to become more selective with the students they admit. The reason for increased selectivity is simple: accepting students who score higher on state reading and math scores is more desirable for high-achieving schools because it better ensures favorable assessment scores and larger types of investment and compensation from the state. Such a market approach reform strategy based on a competitive enrollment system, in other words, "create[s] an educational marketplace that presses schools to compete for students. This is good for the students selected for the strongest schools but not so good for the children left behind and grouped as the weakest" (Winerip 2011). Not surprisingly, the children who are left out overwhelmingly qualify for free and reduced lunch and come from a home where English is spoken as the second language. Here, the market is shaping at least two distinct educational populations: one that has a richer potential for "useful" human capital and another that is, what Henry Giroux and others have called, "disposable" (Fine et al. 2004; Giroux 2010, 2012). In this sense, the value-added movement must not be viewed as solely a method that makes the individual student his/her own "control variable" for performance review. Rather, the other biopolitical feature of value-added strategies is the shaping and regulating of the educational population in the United States into unequal sites of investment *and* disinvestment—the letting live of some and the letting die of others.

Value-Added Schooling and Bio-inequality

What the current debate on whether the value-added approach is an appropriate tool for evaluating student learning and teacher performance fails to take into account is an understanding of how educational life has increasingly become the target of neoliberal forms of governance. Value-added metrics, as I have argued above, are one of the clearest examples of a neoliberal technology of control used by the governing structures of states to regulate educational life within the population. Within the terrain of neoliberal educational restructuring in the United States where *Homo economicus* is the model student, value-added metrics play a particularly important role. They are the latest in a series of educational technologies of measurement that work to normalize the view in individuals and the general population that educational life is something to be optimized through the subjectification of market habits and behaviors. Yet as I have pointed out in my analysis of Fogel and Engerman's work in the previous chapter and the value-added movement here, neoliberal strategies utilized for restructuring educational spaces draw on a variety of governmental technologies to control and regulate both populations and individuals in US schools. Human capital metrics are thus part of a larger system of neoliberal forms of governmentality (merit pay contracts for teachers, a de-regulated student loan industry, open enrollment in school districts that allow for white flight from "under-achieving schools," charter school system expansion, and so on) that have as a goal to produce languages and practices of daily life geared toward a spectrum of optimization on one end and disposability on the other.

The debate around how to solve the ongoing legacies of social and economic inequity through schooling (such as the growing achievement gap or drop out rates between white and non-white students) at closer look increasingly resembles the picture painted by Fogel and Engerman in their attempt to rewrite the experience of slave life within the plantation system. Once *Homo economicus* has been established as the genetic origin of all human life, governmentalities infused with human capital rationality attempt to organize people and groups in society along hierarchical and exclusionary terms, though perhaps not as explicitly as it did in times of legal slavery and the attendant eugenic policies of social and political ordering that maintained such a system. Resegregation understood as a problem of human capital production to be measured by value-added metrics, for example, recasts structural forms of racism as evolutionary outliers to the natural progression of market growth. Longitudinal forms of sociocultural oppression as a result are recoded through different means, while the results of endemic inequality produced through the public school system have become even more pronounced.

Here, we have arrived at a distinctly biopolitical moment in my analysis of human capital metrics. Neoliberal governing strategies such as cliometrics and value-added metrics develop out of globally competitive economic structures that also co-articulate with colonial and white supremacist power structures that have historically undergirded modern states (Foucault 2003). As such, human capital metrics exemplify how forms of inequity are actively managed through governing technologies that are geared toward normalizing human capital practices and discourses in both the educational subject and administrative policies of the state. What I am suggesting makes value-added metrics an example of such a technology of control, and thus biopolitical in nature is the role they play in regulating inequality in schools by expanding the way divisions in the educational population are made through more subtle and nuanced practices. Not unlike the legally segregated schooling era where divisions between white and nonwhite schools were sanctioned outwardly by the state, value-added metrics serve as a device to divide the educational population into different types of human capital stock, though in a dicedly "colorblind" fashion. In other words, value-added metrics continue racial segregation within society but not quite the same way that defined schooling in the eugenic era in the early part of the twentieth century which rested on scientific divisions in the population of superior and inferior subspecies of humans. In a society organized around human capital accumulation, racial difference is no longer a problem of population degeneracy or impurity per se; it is a problem to be dealt with through governing technologies designed to create the highest yield of human capital in the population, the ultimate goal of value-added metrics. What value-added metrics and human capital social policies in general perpetuate from earlier eugenic policies is the "introducing [of] a break into the domain of life that is under power's control: the break between what must live and what must die. The appearance within the biological continuum of the human race of races, the distinction among races, the hierarchy of races, the fact that certain races are described as good and that others, in contrast, are described as inferior: all this is a way of fragmenting the field of the biological that power controls. It is a way of separating out the groups that exist within a population" (Foucault 2003, 254–255).

One of the clearest ways human capital metrics maintain racial divisions within the population that had previously been legitimized through racial science in biological terms, is the manner in which race in the value-added framework is rendered into a nonfactor or as distorting data in the process of determining the educational value of students, teachers, and schools. Currently the public education system in the United States is producing historic levels of social segregation, and in some large urban areas such as the Southern California region, "two out of five Latino students and nearly one-third of all African American students in the region are enrolled in intensely segregated learning environments—schools where 90–100% of students were from under-represented minority backgrounds" (Orfield et al. 2011, 7). While most human capital-based reform approaches in education claim to be a direct challenge to historical forms of racial inequity schools reproduce in the United States, they in fact reinscribe the racial power relations of schools organized to benefit and maintain white supremacy and the capitalist productive paradigm that requires divisions among groups and individuals in order to succeed.[11]

One thing I want to be clear about with this proto-eugenic aspect of value-added approaches is how human capital metrics, as a tool used to determine the value of educational life in the population, are invested with a type of "power that has the right of life and death, wishes to work with the instruments, mechanisms, and technology of normalization" and therefore "it too must become racist" (Foucault 2003, 256). Yet as Foucault made clear, modern forms of biopower used by the state do not necessarily only include the physical killing of individuals and groups in the population. Educational biopower embedded in human capital models of reform such as value-added strategies also encompasses "every form of indirect murder: the fact of exposing someone to death, increasing the risk of death for some people, or, quite simply political death, expulsion, rejection, and so on" (Foucault 2003, 256). Value-added metrics, as a neoliberal solution to the question of racial inequality that has plagued public schooling in the United States its entire history can equally be understood as a device that increases the risk of social death for some while privileging others through greater access to human capital resources and less exposure to life-diminishing environments. The problem of resegregation, again, demonstrates how human capital metrics only maintain racial and class divisions in the population through market approaches to educational reform that are increasingly geared toward the screening and sorting of human capital in society. Recall that one of value-added metrics' greatest virtues, according to its creators and supporters, is that race, home culture, family income, gender, and other "contaminants" are not considered in the measurement of the teacher–student relationship. In fact, race and class, the two most important indicators of whether students attend highly segregated schools for example, are not the focus of reform strategies utilizing value-added approaches. Rather, value-added metrics focus on adjusting the teacher's ability to maximize student's human capital accumulation through a variety of market incentives such as merit pay and teacher rating scores, and in many cases, publishing these scores as an act of public shaming. Human capital approaches to problems such as resegregation reflect the fact that in neoliberal societies, one of the central concerns about the education of its citizens and communities hinges on the question of how to optimize its human capital stock. As such, it is a society that has already accepted a type of biopower that "has the right to life and death" and, by extension, some life is deemed worthy of making live while others are not.

However, Foucault's analysis of how biopower operates in society (of which value-added discourses and practices are a prime example) fails to fully elaborate on how exactly something like school resegregation patterns and human capital production reform strategies are part of a biopolitics that produce death, expulsion, or rejection for many groups within a state's population. In his work that studies how the production of social inequity is fundamental to Foucault's understanding of biopolitics, Didier Fassin emphasizes an important and underdeveloped aspect to Foucault's original schema that I think helps elucidate the biopolitical nature of human capital reform strategies. Fassin suggests that "biopolitics is not merely a politics of population but is about life and more specifically about inequalities in life which we could call bio-inequalities . . . it is about not only normalizing people's lives, but also deciding the sort of life people may or may not live" (Fassin 2009, 49). In clarifying the distinction of how a biopolitics of "letting live and letting die" is defined in contemporary societies, Fassin draws upon the example of apartheid in South Africa. Here, Fassin makes the persuasive point that apartheid was not necessarily a matter of rendering individuals and populations as socially dead (as Giorgio Agamben's concept of *zoë* suggests, for example), but rather it is the type of life that is created or not created through biopolitical models of governance. Thus "political and social choices in labour, housing, education, justice, welfare are made every day, which have immediate or long-term consequences in terms of making or unmaking inequalities of life and acknowledging or dissimulating them" (Fassin 2009, 56). It is such a biopolitics that I am suggesting both Fogel and Engerman's model of social learning attributed to the plantation as well as value-added approaches are ultimately enmeshed. What value-added metrics represent, therefore, is a configuration of educational biopolitics where life is measured by the addition or subtraction of economic (and thus social) value to a student's life while other non-vital forms of life that threaten or are of nonvalue to the existing power relations associated with neoliberal forms of governmentality are exposed to what Achille Mbembe has called a type of necropolitics where "new and unique forms of social existence in which vast populations are subjected to conditions of life conferring upon them the status of *living dead*" (Mbembe 2003, 40).

Another implicit layer therefore built into Fogel and Engerman's human capital analysis of plantation life and particularly its beneficial learning relationship with free-market forces is the claim that human life is enhanced in its entirety through the gradual subjectification of free-market habits and desires both within individuals and populations. In other words, the life expectancy of slaves in Fogel and Engerman's story is increased only through the subjects' acceptance and performance of free-market ethics and behaviors that are a result of the establishment of forms of governance in society that compel such subjectivities into action. Here, the educational impact of laissez-faire social systems of learning retains a distinct biopoltical dimension: educational life is reconfigured as a part of an ordering of social and political life based on one's ability to become an investing and consumptive subject of education. One's worth as a political subject, what individuals and populations are evaluated upon therefore in the neoliberal schooled society, is the amount of value he/she adds to not only the political economy of a society but also to a definition of political life that reestablishes the question of inequality through neoliberal regimes of governmentality.

Value-added metrics are the most advanced example of such a type of governmentality associated with neoliberal schooling. The practices of merit pay and teacher performance accountability in general are a direct intervention into the daily activity of teachers and students' lives—it shapes the social environment in which education is taking place by compelling both teacher and student to act and behave as investible units if they want to achieve academic success (and ultimately healthier, longer lives). Those who choose not to participate or resist the process of human capital accumulation are jettisoned to other institutional spaces and technologies of control such as special education programs, alternative schools, the military, prison industrial complex, or to the augmenting drop-out pipeline of African American and Latino/a students.[12] At both the level of discourse and theory, value-added approaches are colorblind metrics in that they create contexts for measuring educational value by expunging factors such as race, family circumstance, poverty levels, and cultural experience. Yet through governmental strategies such as value-added techniques, racial inequality is reinscribed through the language of the free market, while the very same populations that suffered the most from modern (biological) forms of racialized schooling (separate but "equal") are still the ones that are the object of biopolitical regimes designed to create and maintain "bio-inequalities" in society. What needs to be focused on I am suggesting, are what Fassin has identified as "bio-inequalities"—the reemergence of racial sorting through measurement and assessment techniques, but in a characteristically neoliberal form. Value-added schooling, when looked at from the perspective of bio-inequality production, I am arguing, should be understood as producing inequitable forms of life along a spectrum where, on the one end, those who embody the neoliberal ethic of self-care (an entrepreneurial, self-investing being) enjoy a greater probability to live longer, accumulate more wealth, and have greater access to high quality health and educational institutions. The production sites for these types of bio-inequalities, for example, will overwhelmingly be (and are) in school districts with higher levels of investments, predominantly white, and backed by affluent communities. On the other end of the spectrum bio-inequalities are produced in sites governed through deep social disinvestment, where human capital investment opportunities are so limited or non-existant that greater probability toward shorter lives, incarceration, military enrolment, less wealth, diminished dietary and health options, become the most reliable outcome of institutions such as uraban schools.

Thought of differently, the bio-inequalities being produced through an extractive relation to educational populations associated with neoliberal governing strategies such as value-added metrics, can also be characterized as a type of biometric imprinting: instead of biological forms of racism, bio-inequalities are not measured by genetics markers per se (what the science of biometrics is concerned with) but by the ability of individuals and populations to absorb value-added inputs into their subjectivities in a measurable and calculable way. Social health and sickness, which are directly tied to an individual's ability to increase his/her income-earning potential over a life span through human capital investments, just as it is in Fogel and Engerman's cliometric universe, are bracketed within the economic grammar of neoliberal life. It is not about getting rid of inequalities such as massive resegregation in urban and suburban school districts; it is about managing inequality in a way that best optimizes neoliberal needs for increasing the country's human capital stock while sorting those who are not desirable by determining that they have too many negative forms of human capital to be rational investments. Non-investible students (externalities) instead end up in low human capital regulatory institutions such as prisons, the military, GED programs, or underfunded and overcrowded urban schools. Viewed from this perspective, value-added metrics are best understood as tools designed to measure individuals' and groups' ability to incorporate the genome of *Homo economicus* more and more into their daily lives.

What I am ultimately arguing is that the simultaneous production of bio-inequalities and optimizable life is a deeply rooted mechanism within the human capital framework of education. It should not be forgotten that human capital theory was developed and tested in studies such as Fogel and Engerman's during the ascension of neoliberal thought that moved from margin to center in the late 1970s. Other human capital theorists such as Theodor Schultz who sought to develop ways to measure the quality of populations in poor countries throughout his entire career also reflect this characteristically biopolitical project underlying human capital explanations of life. One such example Schultz draws in his population study of human capital in India holds together two important biopolitical tensions I want to emphasize in order to provide a more nuanced understanding of how bio-inequalities are produced through human capital practices and discourses. Specifically, I want to emphasize a deeply held desire within the human capital project to manipulate genetic structures and environments of life that characterize the ethic of control and domination ensconced within neoliberal practices of governmentality.

It is not a coincidence that the most common stage used to rehearse human capital scenarios was agrarian; the plantation, developing world agricultural settings, and rural-impoverished life in general were (and remain to be) the default proving grounds for human capital theorist to bring to life *Homo economicus* and the world he inhabits. There are two reasons I see the agronomist theme as an important one for understanding how bio-inequalities are constitutive to the human capital project. First, agrarian settings give human capital theorist a "state of nature" to use as a foil for creating their free-market cosmos. Not unlike classic liberal social contract theorist such as Hobbes, Locke, and Rousseau, human capital theorists also needed an artificial construct of nature to build their argument that offered an explanation as to why individuals would want to voluntarily join society and establish a government to protect their lives and property. Agrarian settings also play this role for human capital theorists but with a decidedly neoliberal twist: a natural condition in which to prove the governing laws of nature and human society is universally free market in origin. If reason lies at the heart of such natural laws for Locke and company, it is purely an economic rationality for Schultz, Fogel, Engerman, and so on. Here, nature (in the setting of the farm) and human nature are fused in a model of development indebted to both free-market progress and technological domination over nature.

The second reason agrarian contexts proved indispensible to human capital theorists is because the farm is where people in "low-income countries do all they can to augment their production. What happens to these farmers is of no concern to the sun, or to the earth, or to the behavior of the monsoons and the winds that sweep the face of the earth; farmers' crops are in constant danger of being devoured by insects and pests: Nature is host to thousands of species that are hostile to the endeavors of farmers. Nature, however, can be subdued by knowledge and human abilities" (Schultz 1981, 17). Here, in humanity's ability to dominate, make more efficient, and extract value from nature a core value system of human capital theory is apparent. Yet it is not just plants, soil, fertilizer, and irrigation systems that human capital theorists such as Schultz are interested in analyzing on the farm. More important perhaps is the type of human that is produced alongside the modernization of agriculture practices in developing world contexts. In other words, for both the crop and human capital stock of a population to increase in quality and efficiency, a type of genetic intervention must occur: one at the molecular level of the plant, the other in the social gene pool of a population. In discussing the dynamic between growth in a population's human capital and agricultural output, Schultz (1981) offers this educative example:

> The process of development in many low-income countries is benefiting appreciably from indigenous university trained professionals in engineering, technology, medicine, public and private administrative work, and agriculture. The high-yielding Mexican wheat had, for example, several genetic limitations under Indian conditions. In a few years Indian agricultural scientists modified and improved the genetic composition of the Mexican wheat appreciably, and in doing so reduced these limitations. This achievement on the part of Indian scientists was possible because India already had a sizable stock of this specialized scientific ability. Universities in India had contributed substantially to the training of her scientists in these skills. (46)

Embedded in Schultz's celebratory account of the confluence between human ability highly invested in scientific and technological expertise (one of the most valuable forms of human capital) and the transformation of a plant's genetic structure is a disturbing feature that animates human capital studies of social contexts. In the above example that highlights the human triumph over a type of wheat plant through human capital investments in agricultural science and genetic engineering in this case, we can see that one of the ultimate goals of neoliberal thought is to intervene into the life structures of biocapital: the plant and human scientist. Yet what cannot be ignored are the bio-inequalities produced through human capital strategies of development that treat dietary and human capital impoverishment in India as a problem of crop and population stock quality. On the one hand, we see that local and place-based agricultural practices and knowledge are considered non-vital resources that must die away in order for market developmental models to take hold and flourish. In fact, the problem of poverty in India for Schultz and other human capital theorists can largely be attributed to indigenous and "backward" knowledge and cultural practices. On the other hand, the stock of human capital in India must also be subjected to a radical liberalization of its economic structures so that the proper degree of scientific and technological know-how can be achieved. Here, the concomitant bio-inequalities, again, take form through the erasure of cultural experience and knowledge and the artificial substitution of market-dependent approaches to life in the field and school. What one could argue, and is in fact what I am suggesting, is that human capital theory ultimately calls for a genetic manipulation of the biological resources available in socio-political settings. In other words, both the population of India and the wheat plant required human capital development programs to address genetic deficiencies: one in the genetic structures of the plant and the other in the social genetic development of the population.

Getting Genes Right: Establishing a Genetic Basis for Quality Human Capital in Schools

Fogel and Engerman's study of slave plantations and the value-added assessment reform movement, as I have tried to argue in this chapter, reflect the way human capital theory understands educational life as a zone to be regulated through market rationalities and attendant governmental structures that attempt to create the most value from society's human capital stock. In an increasingly biocapitalist organized society that is driving market expansion into more and more areas of non-commodified life, however, human capital frameworks have taken on even greater significance. Human capital tools such as value-added metrics display a disturbing congruency with the productive paradigm of biocapitalism that is built on rendering market value from biological sources, most notably by reconfiguring the genetic structures of humans and nonhumans to create wholly new products. In Schultz's example of neoliberal agricultural reform in India, we can see that educational and agrarian life have been fused together into the same biological field of control and regulation. Both areas of growth (the field and educational institutions) in Schultz's study on population quality are concerned, for instance, with increasing the yield of valuable biological stock, developing revolutionary methods for making life more efficient and valuable, reconfiguring biodiversity into "pure" strains (cultural and historical experience from the classroom and indigenous plants and farming practices from the field), and establishing models of economic development that are tied to neoliberal restructuring projects rooted in a competitive model of limitless global economic growth (the World Bank and IMF for rural India, and sweeping federal and state reform policies such as Race to the Top). It is not a random coincidence that the founder of value-added assessment metrics is trained as an agricultural bioengineer and that Chicago School human capital theorists found the farm such an appealing experimental site for neoliberal theory and policy development.

What I want to emphasize from my analysis of human capital theory and tools of measurement I examined in this and the previous chapter is the intensifying forms of biopolitical control that are coming to decisively shape educational reform in the neoliberal era. One of the clearest ways to see how technologies aimed at the control and regulation of educational life are playing an increasingly significant role in the structures of schooling in the United States is to understand human capital approaches to educational reform within the paradigm of biocapitalism. That is, schooling as articulated through human capital rationalities and practices mirrors the productive engine of biocapitalism built on taking existing biological material, applying extractive technologies of measurement and sorting, and ending up with an optimized and thus more valuable unit of biocapital. As a way to make the relation between human capital theories and practices of education and biocapitalism more apparent, I will sketch a preliminary map of how value-added metrics dovetail with recent behavioral genetic research attempting to establish a genetic basis for intelligence and new ways for adjusting the conditions for intelligence to emerge in places like the school. In the genetic search for human intelligence and its relation to educational achievement, human capital desires and biocapital modes of production achieve full symmetry.

Researchers' quest to locate the genetic origin of human intelligence has been ongoing for over a century.[13] In the eugenic age of the late nineteenth and early twentieth century, the scientific paradigm placed human intelligence in relation to the subspecies group a person belonged: Aryans held the highest degree of intelligence, while Mongoloids retained the lowest. In the post-eugenic age (after World War II when modern Western science largely abandoned eugenic explanations of race), however, discovering a genetic origin for human intelligence has carried on with great vigor. In the field of education, multiple explosive debates have erupted between genetic researchers' claims that intelligence is largely a matter of hereditability while the effects of environmental factors such as home culture or socioeconomic status are negligible. From Arthur Jensen's (1969) article "How Much Can We Boost IQ and Scholastic Achievement?" that the *Harvard Educational Review* decided was important enough to dedicate an entire issue, to Richard Herrnstein and Charles Murray's (1996) infamous *Bell Curve*, researchers looking at differences in educational achievement between groups of people in society have long been using new findings from genetic science to make arguments that intelligence is largely a matter of hereditability.[14]

Perhaps the most well-known contemporary researcher in behavioral genetics working in the area of human intelligence and educational achievement is Robert Plomin. Trained in the field of psychology and currently working in King's College of London's Social, Genetic, and Developmental Psychiatry Centre, Plomin has spent a good deal of his career studying and researching genetic connections to the development of intelligence in humans.[15] One of the more important implications of Plomin and his colleague's work has been the shift in the grounds of behavioral genetic debates it has caused. Primarily, Plomin and other researchers in behavioral genetics have precipitated a paradigmatic change in how researchers understand the origin of intelligence by reducing the importance of environmental (or nurture) factors and aggrandizing the role of genes (nature) in the development of social and psychological behavior. The shift from a more equal account of the role nature (environment) plays in human intelligence to a heavily genetic one, among other important implications, points more assertively toward behavioral intervention strategies through what Plomin and others call "bottom-up" approaches in molecular biology (such as gene manipulation, gene expression profiling, and proteomics [the study of proteins]). Bottom-up approaches in molecular biology thus call for an approach to human behavior adjustment that operates from a rationality seeking to answer "how genetic effects interact and correlate with experience, how genetic effects on behavior contribute to change and continuity in development, and how genetic effects contribute to overlap between traits . . . these issues are central to quantitative genetic analyses. Behavioral genomic research using DNA will provide shaper scalpels to dissect these issues with greater precision" (Plomin and Spinath 2004, 125). Different from mainstream genetic research on intelligence which holds that gene development and environment interacted on a two-way street, Plomin asserts that "correlations between DNA differences and behavioral differences can be interpreted causally: DNA differences can cause the behavioral differences but not the other way around" (ibid., 119). What Plomin and other researchers doing similar work in the field of behavioral genetics have effectively transformed is the causal scope of genetics in the development of human intelligence. That is, according to Plomin, environmental factors such as family environment are overwhelmingly explainable through genetic predispositions, giving genes the final say on complex sets of relations between environment and DNA that constitute human intelligence.

Indeed, for Plomin and others in the field of behavioral genetics, "one of the most exciting directions for genetic research on intelligence is to harness the power of the Human Genome Project to begin to identify specific genes responsible for the heritability of intelligence. It should be noted that DNA variation has a unique causal status in explaining behavior" (Ibid.). One of the ways Plomin and his colleagues have applied behavioral genetic research to education has been to draw genetic connections to intelligence beyond children's hardwired "IQ genes." In a recent study conducted by Plomin and other behavioral genetic researchers, for instance, evidence was presented that suggested student's self-perceived abilities (SPAs), the other top predictor of school achievement next to student IQ, are 50 percent determined genetically. In essence, what such a study is attempting to establish is a genetic foundation for understanding how students think about themselves and their capabilities and potential in the classroom. In studying groups of twin children (2,287) in the United Kingdom, Plomin and the other researchers working on the study concluded that "contrary to extant theories, SPAs are substantially influenced by genetic factors, and they are influenced by genetic factors at least as much as IQ is." The authors go on to claim that the "results indicated that a common set of genes affected all three constructs, not only when they were assessed contemporaneously (at age 9), but also when each construct was assessed at a different age (IQ at age 7, SPAs at age 9, and achievement at age 10). These results provide evidence for temporally stable genetic influences on the three constructs" (Greven et al. 2009, 760).

So why does it matter that Plomin and other behavioral genetic researchers claim that students' self-perception of themselves, their own subjective attitudes toward their potential as effective learners, combined with independent IQ genes is, at minimum, 80 percent genetically determined (Greven et al. 2009)? And perhaps more importantly to my examination of human capital in this and the last chapter, how does a model of human intelligence based so heavily on genetics connect to measurement tools such as value-added metrics?

The best approach in which to see how value-added metrics and genetic explanations of intelligence co-articulate with one another is to look at how well the value-added measurement device fits with genetic understandings of students' bodies and the abilities they may or may not retain. One way to illuminate specifically how value-added metrics and genetic models of learning work in unison is to understand value-added measurement as a biometric device. What I mean by biometric is simply the practice of collecting biological data from human bodies (students) that form indices that are utilized to measure, sort, and regulate biological material within an environment (school). If, as I have argued, students under human capital types of governmentality can most accurately be understood as units of biocapital, then understanding the genetic structures that lead to desirable behaviors and abilities such as high intelligence would be an invaluable source of knowledge to many neoliberal reformers. The reason such a genetic discovery would be so important to states and governments interested in increasing yields of quality human capital in their populations is because identifying those bodies that carry the proper genetic coding for high intelligence through biometric devices such as value-added metrics allows for a greater degree of efficiency in determining how to invest in which types of biocapital. In other words, the existence of a genetic map of both IQ and student self-perception attitudes lends itself seamlessly with measurement devices that are already calibrated to eliminate environmental factors such as family background, socioeconomic status, or social constructs of race, sexuality, and gender which all affect the learning experience of students and focus on rating and quantifying the amount of value produced in the teacher–student relationship. Measuring for human capital value in students and the population becomes much easier once the genetic codes of intelligence have been established.

In figure 2.1, I have created a diagram to show how I see the synergy between genetic learning models and value-added metrics unfolding in the biocapitalist age. Moving from left to right, we can see that the starting point of student populations (the raw material of educational biocapital production) is filtered through value-added metrics that sort and measure the educational potential of the population's biological material. Next, students are fragmented into varying degrees of high- to low-yield genetic potential, which is required in order

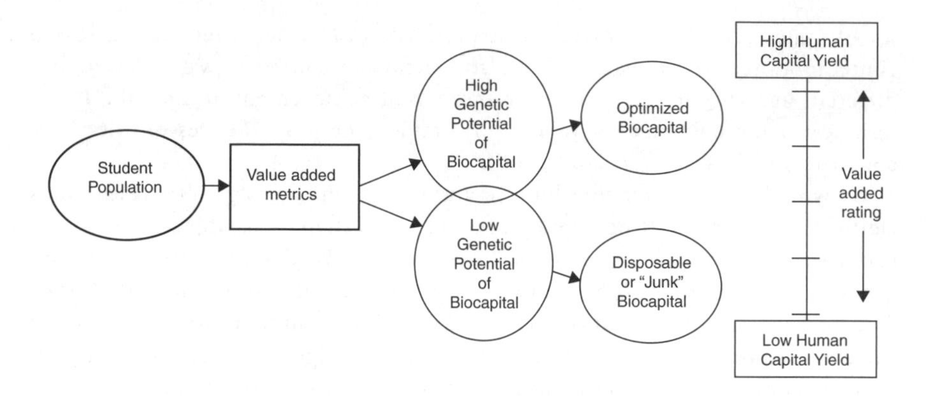

Figure 2.1 Value added metrics as biometric device.

to maximize the value of existing educational biocapital. After the student population has been strained into segments ranging from high to low human capital stock based on genetic potential, smaller groups of biocapital are formed that are invested according to genetic predispositions toward valuable types of human capital such as high intelligence, while other population groups are disinvested in and removed to alternative routes designated for less valuable biocaptial stock. Ultimately, as the diagram suggests, students as units of biocapital are rated on a continuum of value rich to value poor that is arrived at through the combination of genetic knowledge and metrics designed to optimize the potential value of the student population.

What the above diagram suggests is not some far off potential future. Education in the age of biocapital will require that we recognize how existing bioscientific and biotechnological rationalities and practices connect with and extend neoliberal strategies of governance in society. Some relationships such as behavioral genetics and human capital-driven education reform undeniably point to the emerging landscape of education in a biocapitalist society where students' genes become part of the calculable field of biocapital production in schools. If the stakes have indeed risen to where the very building blocks of life have become involved in maintaining types of inequality and the insatiable search for more sites of commodifiable life, how do we begin envisioning and practicing an alternative biopolitics of education? Perhaps one place to begin is to see how scientific literacy, the way science and technology are to be understood in schools in the biocapitalist era, has become central to the promissory futures biocapitalism portends.

PART II

Promissory Future(s):
Learning the Science of Life

CHAPTER 3

Engineering Promissory Future(s): Rethinking Scientific Literacy in the Era of Biocapitalism

> The future man, whom the scientists tell us they will produce in no more than a hundred years, seems to be possessed by a rebellion against human existence as it has been given, a free gift from nowhere (secularly speaking), which he wishes to exchange, as it were, for something he made himself. There is no reason to doubt our abilities to accomplish such an exchange, just as there is no reason to doubt our present ability to destroy all organic life on earth.
>
> Hannah Arendt

Introduction: Educational Life and Biocapital

Hi Kids,
Welcome to the Biotechnology Basics Activity Book. This is an activity book for young people like you about biotechnology—a really neat topic. Why is it such a neat topic? Because biotechnology is helping to improve the health of the earth and the people who call it home. In this book, you will take a closer look at biotechnology. You will see that biotechnology is being used to figure out how to: 1) grow more food; 2) help the environment; and 3) grow more nutritious food that improves our health. As you work through the puzzles in this book, you will learn more about biotechnology and all of the wonderful ways it can help people live better lives in a healthier world. Have fun!

The previous paragraph is an excerpt from the introduction to the Council for Biotechnology Information's (CBI) biotechnology workbook for children entitled *Look Closer at Biotechnology.* The workbook is designed as a lesson plan on biotechnology to be used in science classrooms across the country (CBI 2012). The CBI, a public relations front group for biotechnology industry leaders BASF, Bayer CropScience, Dow AgroSciences LLC, Du Pont, Monsanto, and Syngenta, on the surface created the workbook to teach children about the panoply of brave new food products and health benefits the promissory abilities of biotechnology can deliver to the citizens of the world. Yet, underneath CBI's educational message that valorizes how the bioagricultural industry reconfigures existing forms of life into commodifiable wares, all in the name of human and planetary health, a more subtle lesson is also being peddled. What the CBI lesson plan teaches and other educational enterprises operated by biotechnological corporations such as the Biotechnology Institute (a front group for the biomedical and biopharmaceutical industry) that provides teaching resources and runs the annual BioGENEius competition for students is a particular type of salvationary vision that encapsulates the promissory ethos of biocapitalism.[1]

The definition of biotechnology offered by CBI, for instance, reflects the salvationary and promissory tropes of biocapitalism quite well: "Biotechnology then, is a tool that uses biology to make new products. For example, agricultural biotechnology is a precise way to make seeds with special qualities. These seeds will allow farmers to grow plants that are more nutritious, more resistant to pests and more productive. Biotechnology is a tool for looking closer at nature to find solutions that improve the health of the Earth and its people" (CBI 2012, 2). On the one hand, biotechnology is presented as a divine tool capable of transforming ordinary items found in nature into beneficial new products that enhance human life across the globe. On the other hand, the CBI's definition also portends a speculative element: nature is full of profitable secrets that need to be decoded and turned into something useful and valuable for humankind. The only thing standing in the way of discovering the mysteries of nature is the knowledge and practices of technoscience that the CBI workbook, as an educational device, is seeking to produce more of from the educational system in the United States. In this sense, we cannot think of something like the CBI children's workbook outside of the context of the economic needs of biocapitalism and by extension the neo-*Sputnik* direction of educational reform in the United States that forefronts science, technology, engineering, and math (STEM) as both the salvationary and promissory future of education.

Resting at the intersection of biocapitalism and the neo-*Sputnik* science education reform movement in the United States is a deeply troubling biopolitical project which this chapter attempts to map.[2] Emerging from the promissory landscape of biocapitalist forms of education is an educational subject that is being shaped through a developmental model that involves life in circuits of exchange and valuation in disturbingly novel ways. It is this feature of biocapitalist production, its ethical model that can best be described through its extractive relationship to forms of life, that is the subject of enquiry in this chapter. Reading science education in the age of biocapitalism through a biopolitical lens, I contend, reveals a co-productive project that at once creates subjects who embody an extractive ethic and whose body is also the target of extractive forms of biopower.

As a concept developed within the fields of anthropology of science and science and technology studies, biocapitalism has been theorized as a (re)productive model that arose from the complex set of relations existing between technoscientific research and neoliberal practices of economic development.[3] One of the defining features of biocapitalism is a "mode of generating biocapital

[that] is driven by a form of extraction that involves isolating and mobilizing the primary reproductive agency of specific body parts, particularly cells, in a manner not dissimilar to that by which, as Marx described it, soil plays the 'principle' role in agriculture" (Franklin and Lock 2003, 8). In this chapter, I draw upon such an understanding of biocapitalism that focuses on how forms of life (genes, cells, reproductive organs, chemical compounds of plants, or educational life) have become the target of powerful productive regimes.

The argument I make in this chapter is that in order to obtain a deeper understanding of the biopolitical terrain of high-stakes schooling (where science education is situated prominently), it needs to be investigated how biocapitalism and the set of extractive practices it is built upon as a model of development interfaces with the project aimed at adjusting the future to a past where 'the American Century and the Human Capital Century occurred together' and "follow[ed] directly from the relationships among growth, technology, and education" (Goldin and Katz 2008, 2). The connection between human capital and biocapitalism is particularly important to map since the human capital understanding of education is foundational to the science education reform movement that is oriented toward a co-productive relationship between scientific literacy and biocapitalist economic imperatives. The concept of scientific literacy as it is operationalized in the current neo-*Sputnik* movement, in other words, cannot be understood outside the emergent regime of biocapitalism and the forms of subjectivities needed for its promissory futures to be realized.

The concept of scientific literacy is certainly not a monolithic one, however. In this chapter, I focus on the dominant notion of scientific literacy as it is being constructed through the current science education movement that I argue embodies a biocapitalist ethic. Perhaps the best work being done on more democratic versions of scientific literacy in the field of education is by Wolff-Michael Roth and Angela Calabrese Barton (2004). As opposed to top-down models of science education that impute a traditional form of literacy focused on learning the enquiry process of science and its methods, Roth and Barton ground the development of scientific literacy within participatory contexts that favor a notion of literacy that deals with problems affecting communities directly, such as water pollution, soil contamination, or access to healthy food. Derek Hodson and Nancy Brickhouse's work in science education also strongly forefronts democratic configurations of scientific literacy that challenge racist and sexist values embedded in traditional science education curricula and practice (Hodson 2003, Stanley and Brickhouse 2001, Brickhouse and Kittleson 2006). Others still have done excellent work expanding the narrow traditional understanding of scientific literacy by incorporating views and concepts from research being done in the area of science and technology studies. Here, science educators and theorists have attempted to situate science education within a framework that sees science not as a separate activity from society but instead as a constituent activity that is highly relevant in how we live our everyday lives. For these science educators, the teaching and learning of science should focus enquiry on the ways science and technology affect our communities and everyday understandings of the world around us (Aikenhead and Solomon 1994, Aikenhead 2005).

Finally, the areas of ecojustice and sustainability science education and indigenous science approaches have provided powerful critiques and counter-practices of science education that move well beyond traditional, standardized versions of Western science education that position humans in an antagonistic and destructive relationship to nature (Tippins et al. 2010, Kawagley et al. 1998, Cajete 1999). While the articulation of scientific literacy in the work of these theorists and educators is productive and much needed, none deal with the biopolitical dimensions of scientific literacy as I have taken up here. In particular the ways biocapitalism is shaping science education reform discourses and practices.

In laying out my analysis, I begin by briefly identifying some of the central characteristics of biocapitalism that science and technology studies (STS) theorists have developed in their work. Specifically, I draw upon the interrelated analytics of promissory investment and national salvation for constructing my biopolitical critique of the science education movement in the United States. I choose these categories because they have become a decisive force in shaping the science education reform movement in the United States and, in turn, what the concept of scientific literacy will mean now and in the future. Next, I turn to the clearest example of science education policy that reflects the expanding influence of biocapitalism on the pragmatic development of scientific literacy. The National Academies' *Rising Above the Gathering Storm: Energizing and Employing America for a Brighter Future* (RAGS hereafter) is the latest iteration of a neo-*Sputnik* fervor that is placing intense pressure on educational institutions to develop subjects who are capable of competing in a high-stakes global economy and participating in a nationalistic project for reasserting US economic hegemony—especially in the areas of biotechnological and biomedical research and development. As framed in RAGS, the future security and economic health of the United States hinge on the prospect of increasing the supply of human capital attuned to the types of labor *and* life that the biocapitalist mode of production requires.[4] What is at stake in the science education reform movement, and one important area I explore below, is how the very security of the United States as a global economic superpower, particularly in the area of biotechnological innovation, is a foundational concern driving biocapitalist production. As such, RAGS, and the science education movement in general currently building inertia in the United States must be read within the larger bioeconomic terrain in which many of the most powerful industries and governments in the world have hedged their future.

Yet a large part of understanding the connection between the neo-*Sputnik* fervor of RAGS and biocapitalism is to learn how human capital theories of education are utilized for articulating the pedagogical goals of educational reform today. The human capital model of education, in other words, is pivotal to the success and advancement of biocapitalism and, as such, unpacking the relationship between the two sheds new light on how both are rapidly evolving through a seemingly insatiable desire for subjects highly invested in the skill sets amenable to technoscientific research and development. Clarifying this connection is also important since the notion of an educational subject as a self-investing, entrepreneurial individual enjoys almost unquestioned popular and governmental support. From its early origin in the work of Chicago School economists Jacob Mincer and Gary Becker (the subject of chapters 1 and 2), to *New York Times* editorialist and author Thomas Friedman, to the Obama administration's Department of Education, human capital still remains one of the most influential theories shaping both the individuals' and the publics' understanding of the purpose of education in society today.[5]

In the final section, I begin to sketch an alternative biopolitical model of science literacy that operates from a "politics of vitality" different from the one produced through the human capital model of education prevalent in the biocapitalist era. In developing a provisional framework for an alternative scientific literacy, I turn to the example of the GrowHaus urban farm project. Situated in an underserved and marginalized Denver neighborhood, the GrowHaus represents a political space in which to rethink how the human capital model of education can be reconfigured into a pedagogical device for creating a vital politics that does not operate with an extractive ethic. In other words, I propose that we approach the human capital model in an experimental manner as a way to think through potential alternative biopolitical variations of education that are not tied to an extractive ethic of life but instead one that promotes a healthier and more sustainable form of "biocitizenship." More precisely, in turning to Michael Hardt and Antonio Negri's concept of immaterial labor as a bridge, I want to look at how the ethic of self-investment underlying human capital models that relies on a mastery of immaterial labor can be transformed within learning environments such as the GrowHaus to reverse the power relations currently shaping understandings of life in educational contexts. Such alternative practices of educational investment can be seen as substituting in a different "biovalue" into the "ability machine" that the educational subject has become in the neoliberal moment. Such a reorientation of human capital points to the development of an ethic of biological and ecological health as an alternative model of biocitizenship that can replace the promissory and salvationary model utilized in biocapitalist models of education. Ultimately, what I argue is needed today is an alternative model of biopolitical production of education that is built upon a reconfigured concept of scientific literacy that rejects biocapitalist forms of life and the vital politics it produces, something better attuned to what I term the biodemocratic.

Before I discuss what an alternative biopolitical construction of scientific literacy might look like in educational and social settings, some of the distinguishing features of biocapitalism need to be outlined: specifically, the promissory value framework and salvationary ideology embodied in the spirit of biocapitalist development, as they are integral to the larger analysis being developed here. These prominent features of biocapitalism are important to understand because, as pointed out in the second section, they have been absorbed into the science education reform movement and the concept of scientific literacy it articulates.

Biocapitalism and the Rise of Promissory Pedagogy

One of the principle dynamics energizing the regime of biocapitalism is its propensity to surmount limits to growth in both cultural and natural domains.

Yet unlike Marx and Engel's industrial capitalism, it is not so much new lands, materials, and labor in the traditional sense that biocapitalist actors are targeting as extractive fields, rather it is "life itself."[6] Perhaps the best venue in which to view the circuits of biocapitalism and where STS theorists have focused much of their attention is on the biomedical, biotechnological, and pharmaceutical industries. One of the fundamental goals of biocapitalism can be detected in how "bioeconomic circuits of exchange have as their organizing principle the capturing of the latent value in biological processes, a value that is simultaneously that of human health and that of economic growth" (Rose 2007, 32–33). Here, Nikolas Rose hits on one of the most salient qualities of biocapitalism: its capacity to commodify and make exchangeable vitality in a variety of forms. For example, plant seeds, DNA material, human organs and tissue, information systems integral to processing and reading genetic code, as well as human test subject data are all exchange values within the growing bioeconomy that has arisen in earnest in the past 20 years or so. In making a connection between the biocapitalist propensity to extract latent value from life and the science education reform movement, however, it is necessary to first briefly look at how neoliberalism helped fuse together the enterprise of science with unbridled economic growth, a union that sits at the centre of biocapitalist development.

At the most superficial level, biocapitalism has evolved out of neoliberal economic restructuring that began in the late 1970s in the UK and United States and is generally associated with the removal of restrictive barriers to markets and labor through governmental, military, and corporate intervention. Thus, the privatization of public infrastructure (schools, water, roads, and forests), the individuation of risks (health care, natural disaster responses, waste management, admission and access to higher education, etc.), and the strategic use of military and corporate forces in responding to social and economic crises are all signature qualities of neoliberal governmentality (Harvey 2005, Klein 2007, Foucault 2008 [1978]).[7] Most relevant to this analysis, however, is the elimination of proprietary boundaries in key knowledge producing sites such as the university that became legally sanctioned through the passing of the landmark Bayh–Dole Act of 1980. In effect, the Bayh–Dole Act legalized the partnering of publically funded academic research sites with corporate and governmental entities, which, in turn, opened up new profitable places of production (and investment) in areas such as biomedicine, biotechnology, and biopharmaceutical while completely erasing the line between public/private sites of knowledge production (Haraway 1997, 2007; Kleinman and Vallas 2001; Sunder Rajan 2006; Cooper 2008).[8]

The coupling of academic science departments with the biotechnological and biomedical industries has helped push the neoliberal model of growth into entirely novel regions. Never before in the history of human civilization has an economic system tied so closely the fragile processes of life to a model of development. Vandana Shiva's work on biopiracy, in particular, has been invaluable in mapping out the neocolonial phase of exploitation and proprietary ownership that has arisen with the institutionalization of transnational trade agreements such as the 1993 Trade-related Aspect of Intellectual Property Rights (TRIPS)

(Shiva 1997, 2005). Other governmental/corporate bioprospecting entities such as the International Cooperative Biodiversity Groups (ICBG) have also blurred the line between free-market ideology and academic research in troubling ways as it has been instrumental in "forge[ing] an initiative that would link drug discovery to sustainable development precisely through benefit sharing contracts" that privilege corporations such as Merck and Wyeth-Ayerst while stripping local communities of their ecological and cultural biodiversity (Hayden 2003, 66).

The genealogical feature of biocapitalism most important to the present analysis is how the neoliberal revolution encoded within the sociology of science (and thus reshaping what is needed from the educational system) an insatiable drive to overcome both institutional *and* biological limits to growth. As a consequence, one of the most important effects of the biotech revolution that followed the merging of private and public research was an opening up of "a whole series of legislative and regulatory measures designed to relocate economic production at the genetic, microbial, and cellular level, so that life becomes, literally, annexed within capitalist processes of accumulation" (Cooper 2008, 19). Indeed, one of the most far-reaching effects of the neoliberal project was its driving of economic development in areas of technoscientific research that made it possible to transform biological life into a zone of economic territorialization and commodification. It is this co-constructive relationship between neoliberalism as a developmental strategy and the life sciences that underpins biocapitalism and also contributes to the formation of one of its most pervasive characteristics: its promissory value framework.

In his theoretical mapping of sites of biocapitalist production, Kaushik Sunder Rajan clearly depicts how promissory value permeates private and governmental entities that largely hedge their future on not yet existing goods. In particular, his analysis of biotech start-up labs and the individuals who are involved in their operation and success accentuates how promissory valuation is a key constituent of biocapitalist production. One of the underlying themes Sunder Rajan (2006) points to in biotech productive contexts is an ethic that heavily rests on

> forms of valuation having not to do with the tangible material indicators of successful productivity, but with intangible abstractions, such as the felt possibility of *future* productivity or profit. Vision, hype, and promise . . . are fundamental drivers of this kind of valuation and are central animating factors in drug development, whether it involves the valuations of start-ups by private investors such as venture capitalists, or the valuation of public companies on the stock markets of Wall Street. (18)

Sunder Rajan's point here is that biotech companies, as paradigmatic biocapitalist actors, operate in a speculative environment that drives growth based on a particular vision of investment oriented toward future promises of productive potential, instead of ones based on existing or real value. It is this focus on imaginary productive potential built into the biocapitalist notion of progress that I want to draw out for better understanding the construction of scientific literacy in the contemporary moment.

As indicated above, the promissory ethic embodied in biocapitalist sites of production such as genomic research labs is one made possible through the dynamic relationship established between technoscientific research and biotech industry. As such "the grammar of biocapital is a consequence of the type of capitalism that it is. As a type of high-tech capitalism, biocapital is, certainly in the U.S. context, often *speculative*, a reflection of commercial capitalism almost to the exclusion of commodity capitalism" (Sunder Rajan 2006, 111). Echoing Rajan's insights on the promissory valuation at work in biocapitalist circuits of production, Cori Hayden makes a parallel point in her research on pharmaceutical bioprospecting in Mexico. Embedded in the extractive relationship established between indigenous communities, a US and Mexican University, and a giant pharmaceutical company sanctioned by international governing boards of trade is a type of exchange that is infused with what Hayden identifies as an ideology of "futuricity." As Hayden (2003) points out "in the UNAM-Arizona agreement, as with countless prospecting projects, 'the product' has not (yet) materialized: it is a promise that may remain just that" (74).

It is this speculative ethos driving some of the most powerful areas of the economy that will be important to keep in mind when I examine aspects of science education reform in the neo-*Sputnik* era below. As I argue, much in the same way that investment into imaginary drugs fuel biocapitalist development, a similar promissory framework is present in the construction of educational subjectivities in the biocapitalist era. Thus, one of the most important attributes of the promissory value framework of biocapitalist production informing this investigation is its ability to make valuable something that has not yet come into being. As a result gene therapy drugs, the chemical compounds of plants, and educational subjects, at one level, are all treated as potential exchange values that have been conflated within the imaginary productive zone of the promissory horizon created in biocapitalist society. Yet investing in such a potential future is also driven by a strong faith in the ability of biocapitalist progress to place nations into positions of global economic power. Part of the biocapitalist vision of the future, in other words, also has a lot to do with market and governmental dominance in a highly competitive global economy.

The second feature of biocapitalism being articulated in environments such as the start-up lab that is related to the formation of current science education policy reform is the "risk management" discourse of national salvation. That is, as Sunder Rajan's analysis of genomic research laboratories reveals, biocapital firms that function with a currency of promissory goods tend also to retain a pseudo-religious ideology that links success in technoscientific research to that of the salvation of the nation. It is at this intersection between technoscientific research and national security rendered in the high-stakes biocapitalist context that an important political dimension of science opens up that usually remains obscure, namely, how "the promises of biocapital are undergirded by salvationary and nationalist discourses" which have "salvationary stories . . . embedded in the ethos of specific corporate cultures, and in the cultures of biotechnology and development industries writ large" (Sunder Rajan 2006, 35). Here, it is clear that one of the latent yet constituting features of biocapitalism is the unwavering faith put into its productive framework that brings with it a very potent nationalist discourse linked to notions of security, territorial hegemony, and citizenship. This risk management aspect of biocapitalism plays a significant role in the construction of scientific literacy as it is articulated in the security-oriented science education movement of the neo-*Sputnik* era.

The compression of scientific progress into national security strategies that has taken place in the biocapitalist production paradigm has an important effect on the formation of subjectivities in sites of scientific practice (and by extension science education settings) and is worth considering more closely. An instructive place to look for understanding how such a transfer between national security and technoscientific R&D occurs is biocapitalist sites of production such as the biomedical research lab. The genomic lab is an important model for understanding the salvationary trope of biocapitalism because it represents a zone of subjectivity production that the science education reform movement is seeking to replicate. High pressure biotech labs, in other words, reflect "the co-constitution of highly individualized stories of personal motivation, calling, and human interest with the structural messianism inherent in the market, science or nation," which has the effect of making the act of "saving lives meld into saving a company's corporate interests or, in the case of India, making a 'Third World' country a 'Global Player' " (Sunder Rajan 2006, 198).

What Sunder Rajan's study of both US and Indian biotech actors uncovers is the emergence of a powerful discourse and practice he calls the "born again ethic" of biocapitalism. Thus, in the case of human genome research, a sort of new logic of capital has developed that has a powerful effect on the production of subjectivities according to Sunder Rajan—a logic that is infused with an unfaltering belief in the biosciences and their ability to discover the next ripe market to be exploited for profit and glory of nation. Yet this salvationary ethic also points to a neo-Weberian melding of religious and capitalist values that is also shaping notions of citizenship through the belief system of neoliberal free-market ideology and American exceptionalism. Here, citizenship has more to do with saving one's country through scientific and technological superiority than it does with traditional civic values such as rootedness in community, rational communication and debate between diverse members of society, or the cultivation of these general civic virtues through a system of public education. What emerges from Sunder Rajan's analysis of the biocapitalist productive framework, therefore, is a model of citizenship that is tied to an idea of nation that mirrors the enterprise of high-stakes technoscientific research that takes place in some of the most economically volatile environments in society.

The salvationary character of biocapitalism has also been a large part of Nikolas Rose's work on charting the rise of the bioeconomy as a dominant sector of investment for governments around the globe. For example, Rose (2007) notes that "in 2003 the U.K. House of Commons Trade and Industry Committee Report on Biotechnology identified biotechnology, especially biomedical biotechnology, as a key economic driver and estimated that, in 2002, the U.K. biotechnology industry had a market capitalization of £6.3 billion, accounting for 42 percent of the total market capitalization of European biotechnology" (35). The United Kingdom is not alone, however, as Rose also points out that countries such as China, India, and South Korea also share this salvationary vision in the emergent bioeconomy. For Rose (2007) then, "political investment to support the development of the biotechnology sector in each country and region was driven, at least in part, by fears of the consequences of loosing out in this intense international competition" (36). Given the amount of faith governments and nations have put into the bioeconomy as the most important economic sector, it is not surprising that a corollary notion of citizenship has also arisen in such a salvationary context. Here, success in the labs and markets of the bioeconomy is directly related to the supply stream of highly technoscientifically literate subjects that a country can produce. In such a model, civic duty is largely measured by one's ability to help his/her country compete in "an intense international competition" rather than a democratic political life.

Here, we have reached a key point where educational need and biocapitalism intersect. Coupled to the promissory and salvationary ethos of biocapitalism, as I argue below, is the need for educational subjectivities that retain the proper types of investment most useful to the productive framework of biocapitalism. Scientific literacy and, in particular, how it is being defined in our neo-Sputnik moment are the most important mechanism to view for understanding how the regime of biocapitalism is reshaping education into its own image. Yet key to making sense of how scientific literacy is at the center of such a project it needs to be understood how the theory of human capital underlying science education reform instills the promissory and salvationary ethic as a primary mode of conduct for the educational subject. The relationship developing at the nexus of education and biocapital is one where future-oriented individuals who are perpetually investing in his or her life to increase levels of human capital and participate in a "race to the top" mirror the ethic of the biocapital actors of the lab.

Vitality and Human Capital Education

Human capital education when viewed alongside the promissory value framework of biocapitalism, a project clearly geared toward producing a particular type of educational vitality becomes apparent. The principal problem I want to investigate in this section is how scientific literacy has become the key mechanism for shaping educational life in a way that is congruent with the biocapitalist imperatives of promissory valuation and salvationary discourses. This question is a crucial one as scientific literacy is now more than ever geared toward fulfilling the needs of a rapidly evolving bioeconomy. Not surprisingly, the project for defining scientific literacy is being led by some of the most powerful industries in the United States and across the globe. The pharmaceutical, biomedical, and biotechnological, three of the most prominent sectors of the global economy, have become primary drivers for determining the direction and social future of the US economy. As such, they also have a steep demand for particular kinds of educational labor and skill sets that are reflected in the recent increase in governmental spending patterns over the past decade. For example, US pharmaceutical research and biotechnological companies, despite a historic economic downturn, have invested a record $65.2 billion into new R&D in 2008 alone (PhRMA 2009). On top of such a massive spending, the Obama administration and the US Congress have also allocated $10 billion for scientific research that is tied to the American Recovery and Reinvestment Act. The US government in addition also indicated that 3 percent of the annual GDP will go directly into bolstering efforts to increase scientific and technological research in the United States, Science, Technology, Engineering, Math (STEM) education being one of the most important loci of this legislation (American Recovery and Reinvestment Act 2009).

The need for educational lives attuned to the skills, knowledge, and modes of analyses that are built into the knowledge-hungry bioeconomy places the struggle to define scientific literacy at the heart of economic recovery. As such, both sides of the political spectrum in the United States are pushing for a notion of scientific literacy that can feed directly into the demands created by biocapitalism and the promissory value model it adheres to. What needs to be recognized in this effort is how the desire to manage student populations in a manner that produces individuals who are appropriate for the technoscientific workplace is also part of a larger venture model of investing associated with biocapitalism: increasing investment in STEM education has a high potential to yield dividends in the form of potential knowledge workers. Software developers, data interpreters, biochemical, biophysical, and genetic researchers are the desired labor base of biocapitalist society, and schools, as it turns out, have been targeted as one of the most important promissory production sites.

In this sense, scientific literacy within the biocapitalist context can perhaps best be understood as a strategy focused on the production of educational subjects who retain a particular type of "biovalue" that shares in the promissory ethos of the bioeconomy (Rose 2007).[9] By interpreting educational subjects as forms of biovalue, scientific literacy can more accurately be recast as a technique for managing and regulating educational life as a field of extractable value. One of the most successful ways scientific literacy attempts to achieve its extractive goals is by drawing upon and reconfiguring the human capital ethic that already serves as the basis to the current science education reform movement. By setting scientific literacy within the biocapitalist framework, the organizing field of biopower that normalizes the process of capturing latent value from educational populations becomes apparent. It is on this axis where human capital as a theory and practice of education should be reread as an integral part of the larger goal of science education reform aimed at cultivating forms of life suitable for biocapitalist modes of production. Science education, as conceived within the milieu of biocapitalism, in other words, is a space analogous to one that "can be known and theorized, that can become the field or target of programs that seek to evaluate and increase the power of nations or corporations by acting within and upon that economy. And the bioeconomy has indeed emerged as a governable, and governed, space" (Rose 2007, 33). What I am suggesting requires further examination is how the theory of human capital operating within the science education reform movement participates in a project that has as an aim to increase the biovalue of educational life. One of the best places to view such an extractive logic at work is in the National Academies' legislative recommendation RAGS.

Not unlike the tone of fear that fuelled educational reform during the cold war *Sputnik* era, RAGS has unmistakably placed science and math education within the matrix of national security and economic salvation with a renewed intensity. Building off the Hart-Rudman Commission's flurry of homeland security measures, RAGS embodies the view that "the inadequacies of our system of research and education pose a greater threat to U.S. national security over the next quarter century than any potential conventional war that we might imagine" (RAGS, 25). Remarking on the seriousness of the threat that a declining science and technology research sector represents to the United States, former President George W. Bush has also warned that "science and technology have never been more essential to the defense of the nation and the health of our economy" (RAGS, 25). An obvious question that surfaces in the face of such catastrophic rhetoric, which is not limited to the neoconservative perspective, is what rationale lies behind such a line of thought? One place to look for an answer to this question is to one of the most influential arguments made in constructing the theoretical point of view of RAGS that explicitly ties salvationary narratives and national security concerns to science and math education reform. Here, Thomas Friedman's "quiet crisis" thesis, which hinges on the human capital theory of education, is highly influential to the theoretical foundation of RAGS and thus is worth briefly examining.[10]

Friedman's concept of a quiet crisis is generated from what he argues is the flattening of the economic world through the processes of globalization. Friedman's new flat world is an equalized economic playing field that has been made possible through a combination of information and communication technologies and other countries' greater investment into their educational infrastructure. Yet the new flat economy is nothing to fear, Friedman (2005) asserts, since "American individuals have nothing to worry about from a flat world—provided we roll up our sleeves, be ready to compete, get every individual to think about how he or she upgrades his or her educational skills, and keep investing in the secrets of the American sauce" (252). However, Friedman's hopefulness is also tempered by his simultaneous recognition of US stagnation in key fields of knowledge production when compared to other nations that are quickly surpassing the United States in areas such as biotechnological and biomedical R&D investment. China and India, according to Friedman's analysis, both have work forces that are more driven and better educationally prepared for the competitive flat economy that requires a high degree of technical proficiency, an unbridled work ethic, and innovative thinking. Friedman implores that such a growing accumulation of human capital in other parts of the world should be a wakeup call for US policy makers and corporate leaders alike. If the United States wants to have any chance of retaining the mantel of economic and moral leader of the world, Friedman argues, its educational system must, in short order, remedy some of the structural flaws that are preventing the necessary amount of human capital from being produced. Put simply, the educational population within the United States, according to Friedman's analysis, must be reconditioned for the new coordinates of the flat world—schools and universities being the most important places to begin such a reconstructive project.[11]

What Friedman's argument for a reinvestment in the human capital of the United States's educational population ultimately purports is an equation that is pregnant with contradiction. Following the overall logic of Friedman's "quiet crisis" argument to its resolution, what becomes evident is a comparative choice that is far from desirable. That is, in pointing to countries such as China and India that are investing more into their educational populations as comparative models of productive life, Friedman fails to take into account how these examples are the result of neoliberal pressures that have been a large part of the formation of the landscape of global capitalism. In other words, what Friedman is in effect suggesting is that the laborer in China who works long hours in a massive transnational corporate factory assembling electrical switches or extracting valuable metals from e-waste sites that hold massive amounts of disposed computers and monitors is the model of the motivated and innovative worker of the twenty-first century that the United States should strive to emulate. Here, workers' realities are strictly regimented through the tight organization and disciplining of worker groups in camp-like atmospheres where micromanaging for efficiency, quality control, and the extraction of value from both human and natural life is the dominant ethic.[12] Thus, what Friedman's celebratory framing of flat world production and exchange circuits obscures is how valuable metals attached to control boards in computers and human life are both condensed into a similar form of valuation. In other words, an ethic of extraction rules the model of labor and production that Friedman points to as a competitive model from which the United States should be taking its cue. From a biopolitical perspective, then, Friedman sets up a false choice that promotes a vital politics for the further destruction of life rather than forms of resistance that can lead to an alternative model of politics that jettisons this ethical component that the human capital theory harbors and, in fact, promotes.

Indeed, it is not surprising that Friedman is one of the central theoretical figures informing the RAGS document. As such, one of the top priorities identified in RAGS is to address the problem of how to produce "an educated, innovative, motivated workforce—human capital—...the most precious resource of any country in this new, flat world" that starts by dealing with the "widespread concern about our K-12 science and mathematics education system, the foundation of that human capital in today's global economy" (RAGS, 30). In a strategic sense, RAGS identifies the critical shortage of human capital, or the "disinvestment in the future," as being one of the most urgent problems facing the US economy and its national security. Reading this argument in the light of the promissory ethic that helps animate biocapitalism, the crisis Friedman and by extension RAGS has named can with little effort be reframed into the problem of how to extract greater amounts of human capital from an educational population that lacks the requisite vitality needed for a competitive, technologically and scientifically driven world.

In restating the problem in this way, we are in a better position to view one of the strategies RAGS forefronts for achieving the goal of maximizing human capital production via education, its "10,000 Teachers, 10 Million Minds" program. A plan built around a favorable economy of scale, its design principle rests on the simple formula of recruiting "10,000 science and mathematics teachers by awarding 4-year scholarships and thereby educating 10 million minds" (RAGS, 115).

What is apparent in the 10,000 Teachers, 10 Million Minds approach is a distinct biopolitical project that underlies the science education reform movement writ large. Ranking this program as one of its most urgent, RAGS reveals its promissory framing in formulating the ultimate goal of the 10,000 Teachers, 10 Million Minds program: "Our country appears to have lost sight of the importance of scientific literacy for our citizens... without basic scientific literacy, adults cannot participate effectively in a world increasingly shaped by science and technology," neither will there be a "next generation of scientists and engineers who can address persistent national problems, including national and homeland security, healthcare, the provision of energy, the preservation of the environment, and the growth of the economy" (RAGS, 112). Here, in one of RAGS' top recommendations is the simultaneous deployment of both the promissory valuation and national security tropes. Each, however, needs to be dealt with separately within the RAGS framework in order to fully assess how the project for defining scientific literacy has become fully entwined with the imperatives of biocapitalism: specifically, how the goals of science education have been translated into ones that are in line with developing a type of educational vitality heavily invested with biocapitalist literacies.

Similar to the biotech start-up lab model, the 10,000 Teachers, 10 Million Minds model conceives of schools not unlike the laboratories in which venture capitalists' stake large amounts of capital in the hope of receiving large returns. What should also be recognized in the symmetry drawn here between the lab and school, however, is how the ethic of promissory valuation is shaping the way policy makers and corporate actors frame scientific literacy, and by extension the overall enterprise of public schooling as a site for extracting a more exchangeable biovalue from educational populations. Perhaps one of the clearest examples of the promissory value framework in action is in the 10,000 Teachers, 10 Million Minds real-world model: The Merck Institute for Science Education (MISE).

The merging of science education with one the largest pharmaceutical corporations in the world is not just another instance of increased privatization in the public education system, it also demonstrates how corporate actors have long recognized the importance of investing in education as a strategy for receiving increased value in return. The history of intimacy between educational institutions and scientifically driven industries is a long and well-documented one (Marcuse 1964 [1991], 1972; Noble 1977; Nowotny, Scott, and Gibbons 2004;

Olssen and Peters 2005; Cooper 2009). But in the example of MISE, a new strategic site of production has become apparent. In addition to continuing the practices associated with corporate–academic partnerships that use public funds and spaces for generating private gains, biocapitalism adds to this existing relationship in its targeting teacher development as a promissory site of investment. Here, the public and democratic function of teaching that at minimum is rhetorically connected to the legacy of the democratic public school system in the United States has been completely replaced by a regime of biopower that has reconfigured democracy into something that is largely measured by living capital. By zeroing in on teacher development as a site of investment, the MISE program of the 10,000 Teachers, 10 Million Minds approach only further advances the erosion of public spaces that could allow for alternative critical literacies of science and technology to develop in society. In essence, by affecting the means of production of science education with a promissory framework, what are ensured are more exchangeable literacies and individuals for biocapitalist labor markets.

As a global leader in the pharmaceutical industry that produces products such as Vioxx and Propecia, Merck's active role in shaping science education policy, science education professional development, and standards-based assessment models goes well beyond the pale of simple corporate citizenship. Instead, what we see in the instance of MISE as a proposed exemplary model for transforming scientific literacy into an extension of the logic of promissory investment is a powerful mechanism for shaping educational life. One of the most important ways the MISE teacher education and assessment program transfigures the act of education into a vitality technique is through the epistemic transfer that occurs in a science education context created and managed by biocapitalist actors.

In its effort to replenish scientific and technological human capital reservoirs, Merck does not hide the fact that "fostering the next generation of scientific leaders is a key part of Merck's overall commitment to science: It is essential for the sustainability of our business to have access to the best trained scientific minds globally, and its is essential for the economic development and well-being of the communities in which we operate." Toward this end, Merck has also "provided long-term, sustained support for programs that expand capacity for training in biomedical and health sciences" (Merck, Sharp & Dohme Inc 2009).[13] Residing in the epistemic transfer between a biocapitalist actor (Merck) and science education contexts is a clear example of what science study theorists have identified as "the social" and "the scientific" coproducing one another. As such, the epistemic structure coproduced in the knowledge-producing site of science teacher training program and biocapitalist entity is one that is designed to achieve the goal of "hav[ing] access to the best trained scientific minds." Again, what rises to the surface is how promissory valuation, a principal attribute of biocapitalism, treats students (and teachers) as a strategic site in which to invest and extract particular kinds of biovalue.

Another important effect of the epistemic transfer between biotech industry and science education programs is the complete collapse of biocapitalist values into the milieu of education. By managing science education pipelines for teacher training and development, what is simultaneously foreclosed on is the potential for an alternative future where science education could mean more than what human capital as an educational goal can achieve. In other words, the struggle to determine the productive parameters of scientific literacy in educational contexts is also a struggle to control how individuals treat and understand life in our highly technological and scientific society. What is at stake in the science education reform movement and exemplified in the MISE model, therefore, is an educational life that is defined through biocapitalist progress as opposed to one that calls such a productive model into question. Yet, dislodging the narrative of progress tied to the biocapitalist concept of scientific literacy is difficult because it is also tied to a productive regime of national security strategies.

The second biopolitical aspect of the science education reform movement and its concept of scientific literacy I want to examine is what Foucault saw as one of neoliberal society's most pervasive qualities: risk management strategies. At the most general level, the salvationary trope operating in RAGS grows out of the view that US security in the post-9/11 era is inextricably tied to the endeavor of institutionalizing a mode of scientific literacy compatible with the demands of biocapital. The comingling of risk management strategies and science education stems from political and corporate architects who view science and math education reform as "investing in science (including math and science education) [as] the most important strategic investment we make in continued American leadership economically and militarily" (Gingrich 2006). Helping construct this foreign policy component of science education is the widely held view that biocapitalist industries are the most important sector of the economy and, as such, they cannot be understood outside of national security and market competition matrices. Perhaps, nowhere can the salvationary mission tied to science education reform be seen better than in the American University Union (AUU)'s National Defense Education and Innovation Initiative (NDEII), a strategy that RAGS heralds in its call for initiating security-oriented reform through schooling.

The NDEII is a policy report created by the AAU to call attention to the security threat that a depleted pool of scientific and technological human capital represents to US stability. The coalescing of national security and economic health as it relates to the area of science and math education indeed has a lengthy history in the United States. Understood in this light, the NDEII report falls in historical line in its calling for a second coming of the original National Defense and Education Act that was passed in the aftermath of the *Sputnik* launch in the late 1950s. What is unique this time around, however, is how the ongoing war on terror and economic development driven by biotechnological advancement have collapsed so completely into one another that the project of education has become indiscernible from many of the national security goals of the US government in the post-9/11 era. However, what is of particular interest to this analysis is the way the partnership between education and national security (a naturalized relationship in the reform movement) is given a heightened sense of urgency within the biocapitalist milieu that is shaping the behavior of both educational and military institutions in important ways. The synergy developed between education and security apparatuses in the biocapitalist context is clearly presented in NDEII's report: "The principle ways to secure our nation's prosperity and military capability are to strengthen our educational system and revamp and re-energize the structures for innovation that have served this country so well for the past half century. The concern is clear: if we remain on our present course, our nation will not be able to produce the well trained scientific and technological work force necessary to meet increasing competition in world markets" (Association of American Universities 2006).

The relationship between educational and security projects in a biocapitalist framework such as the NDEII and RAGS is clearly one that is playing out on a more complex plane than past iterations of science and math education reform in the United States. That is, just as the claim that human capital in the areas of science and mathematics education are at threateningly low levels, a similar notion of security can be drawn that equates the ability of governments and corporations to expand zones of production and markets through the acquisition of raw biomaterials throughout the globe. As Cori Hayden has argued in her recent work on bioprospecting, the tension between multinational corporate patent rights, states, and local indigenous communities is only the front line of a larger battle for obtaining raw biomaterials that lucrative US biotechnological industries require to function and meet their future-oriented demands. As Hayden (2003) puts it, "these contracts [bioprospecting rights], which set up the chains of entitlement and access between drug companies and southern resource providers *via* academic scientists, points us to concerns that are not easily contained in the moniker 'commodification' ". In considering this neoimperial aspect of biocapitalism, I am suggesting that any analysis that looks at how biocapitalism is shaping important trends in science education also needs to take into account how the trope of security is very much a two-sided coin: the domestic production of human capital in STEM areas of education and the neoimperial practice of acquiring and privatizing plant materials both share the same aim of extracting biovalue from life. Here, the tethering of education to the project of national security within a biocapitalist model situates science education reform within broader networks of power that constitute the practice of technoscience within neoliberal global contexts.

On the one end of the biopolitical spectrum, as I pointed out earlier, educational life within the matrix of biocapitalism is treated as a strategic site of production. This is largely the case because the development of technoscientific human capital is something that biocapitalism, as a mode of production and exchange, cannot exist without. Yet at the other end, life more generally is also integral to the growth of the bioeconomy, a model that stakeholders in government, the corporate world, and academia have a strong interest in expanding. In such a context, plants in Mexico or human test subjects in India are also part of the formula for keeping the United States secure and globally competitive in the new "flat world" where education plays a key role. Thus, part of realizing how the biopolitics of science and math education operates is being able to interpret the one-dimensional framing of the "quiet crisis" within a larger, multifaceted project that enlists educational subjects not only into productive sites for increasing technoscientific human capital, where individuals themselves are measured as biovalue, but also into participation in a predatory economic model that views forms of life as legitimate zones of production and exchange.

In a Foucauldian sense, I am suggesting that we must think of human capital education operating within the science education reform movement as both a disciplinary and a control technology; the concept of scientific literacy being the most important technique of intervention that is aimed at both shaping individual habits and the general regulation of educational populations. With human capital as the ethical grounding for science education, students are trained into ways of thinking about *and* participating in scientific activity that is also part of a larger project of constructing a promissory future. The imaginary future built into the project of biocapitalism, however, is also dependent on the ability to manage and regulate educational populations by tying nationalistic and imperial goals to science education. With this in view, the biocapitalist project to define scientific literacy in educational contexts today signals the advent of a social milieu that has already "pose[d] itself the problem of the improvement of its human capital in general" and answered "the problem of control, screening, and improvement of the human capital of individuals, as a function of unions and consequent reproduction" (Foucault 2008, 228). If this is indeed the case, and in fact this is what I am arguing, then human capital education, which constitutes the ontological basis of science education reform in the United States, needs to be read within what Nikolas Rose has identified as one of the most prominent features of biopolitics today, namely, how science education is directly linked to the "the marketing powers of the pharmaceutical companies, the regulatory strategies of research ethics, drug licensing bodies committees and bioethics commissions, and, of course, the search for profits and shareholder value that such truths promise" (Rose 2007, 28).

What is missing from the debate surrounding science education reform in the United States therefore, and what my analysis points to, is a recognition of how scientific literacy is an integral component to a larger biocapitalist model geared toward controlling and managing the conditions of educational life in a number of powerful ways. The educational implications to such a way of conceiving scientific literacy are indeed manifold. In the final section, in an attempt to provisionally respond to the biocapitalist concept of scientific literacy examined earlier in this chapter, I focus attention on the question of what an alternative biopolitical model of scientific literacy informed by a non-promissory and salvationary value framework might look like. Put simply, I want to ask from where would a life-affirming model of scientific literacy begin, one that is guided by a notion of "vital politics" concerned with ecological health and social justice as opposed to one that carries with it an ethic oriented toward the capitalization of life.

Scientific Literacy and Biodemocracy

Developing a biodemocratic conception of scientific literacy should start with a strong understanding of the ways in which biocapitalism has fundamentally transformed traditional boundaries between culture and nature, human and nonhuman. As an economic and developmental model, biocapitalism involves culture and nature in countless efforts to enlarge areas of commodification, exchangeability, and consumption. Indeed, as STS theorists have pointed out, the era of politics at the molecular level is now upon us. It is no surprise then that science education has become a strategic target of intervention by corporate and governmental actors who are seeking to reconfigure educational life in ways that will best meet the needs of a fluid and highly competitive bioeconomy. The era of biocapitalism, in other words, presents students, teachers, and citizens with an entirely new terrain in which life is in a state of constant flux and reassembly: mental health, medical treatments and procedures, the genetic manipulation of biological processes, and reproductive technologies all represent a growing constellation of areas of human and natural life mediated by forces operating within the promissory and salvationary calculus of biocapitalism. In such a context, how should we begin to rethink scientific literacy given the fact that it has become one of the most utilized technologies of control for managing social futures in line with the promissory and salvationary ethos of biocapitalism? One point of entry for exploring such a question is to look at the model of citizenship that has emerged in the era of biocapitalism and more precisely, what this type of civic life means to the project of rethinking the intersection of democratic education and biopolitics.

In the binding pact created between scientific, corporate, and governmental and nongovernmental actors that comprise the nexus of biocapital, a unique type of "biocitizenship" has arisen that challenges traditional pedagogical considerations of civic life. As Rose (2007) has noted in his work, "biological citizenship" is one of the most prominent features of twenty-first century politics and signals a new era of vital politics. Rose's concept of biological citizenship (which draws upon Paul Rabinow's theory of biosociality) is based on a theory of subjectivity which conceptualizes individuals as "shap[ing] their relations with themselves in terms of a knowledge of their somatic individuality" while also interpreting how "collectivities [are] formed around a biological conception of a shared identity" (Rose 2007, 134; Rabinow 1996). Rose's notion of somatic politics, or a politics oriented toward the care of the body, is a useful starting point for thinking about the formation of subjectivities in science education environments that are increasingly being shaped through biocapitalist imperatives for a number of reasons.

By utilizing Rose's theory of biocitizenship, the dual formation of individuality and sociality as a point of reference, a pedagogy for civic life embedded in biocapitalist scientific literacy can be elucidated in a productive manner. As Rose has documented in his research on the bioeconomy, many countries are engaged in projects to educate their public about new biotechnological advances in such a way as to inscribe in the individual a sense of personal responsibility for developing biomedical literacies of health. This governmental strategy of placing the onus of collecting knowledge about potential health treatments for one's body on individuals is integral to the process of making biological citizens (Rose 2007). However, such public pedagogies of self-care also rely on top-down models that involve "the creation of persons with a certain kind of relation to themselves" (Rose 2007, 140). For Rose, biological citizenship thus focuses largely on individuals' ability to develop literacies of health, and possible treatment options that are shaped and influenced through a variety of corporate biomedical regimes. Underlying this biocapitalist model of citizenship is a type of civic life that is dependent on biomedical industries and mental health experts for interpreting the somatic lives of individuals.

In applying Rose's notion of biocitizenship to science education contexts, what becomes apparent is how scientific literacy infused with biocaptialist values implicates the body of the individual within not only the consumptive but *also* the productive domain of the bioeconomy. From this standpoint, Rose's theory of somatic politics that underlies his notion of biocitizenship can be extended in viewing science education as a form of biopower that involves the individual as a potential participant (or at minimum a tacit actor) in the very productive regimes defining biological citizenship. Even if individuals do not become active participants directly connected to sites of biocapital production such as biochemical laboratories and genomic research facilities, they nonetheless indirectly accept a concept of citizenship that is involved in the consumptive and pedagogical patterns that the bioeconomy has set into place. In other words, this broader civic function implicit in the biocapitalist expression of scientific literacy normalizes the notion that life is something to be managed and controlled through technoscientific knowledge and practices. Thus, the desire for technoscientific labor underlying science education reform creates both a type of political life (the subject who is educated into the acceptance of the investment/extraction ethic of biocapitalist society) and exchangeable forms of biomaterial (plant compounds, subjects highly invested with technoscientific know-how, genetic information, etc.) as the basis of scientific literacy—the most important tool for regulating public understandings of technoscience in society.

On this point, Antonio Negri and Michael Hardt's biopolitical notion of immaterial labor is instructive. In their reformulation of the traditional Marxist concept of labor, Hardt and Negri persuasively argue that immaterial labor has become the affective skin of biopolitical production in late capitalist society. Data, codes, symbolic interpretation, affective communication, and information and communications technologies all comprise forms of immaterial labor that biopolitical production now largely rests according to their analysis (Lazzarato 1996; Hardt and Negri 2000, 2004, 2009). Reinterpreting Rose's notion of biocitizenship as a model of vital politics alongside Hardt and Negri's notion of immaterial labor, however, helps bring to the surface a striking feature of the neo-*Sputnik* model of scientific literacy: its implicit ontological project of producing educational subjectivities attuned to the circuits of biocapital. Rethought from the standpoint of the production of individuality *and* sociality, what scientific literacy in a biocapitalist framework seeks to create is a pedagogical environment where individuals are indeed somatically defined, but in a way that directly relates their bodies to forms of immaterial labor that fuel biocapitalist production—investment in human capital as a method for cultivating biocapital workers, that is, expands immaterial labor as a form of social power. Similar to Marx's analysis in *Capital* volume 1 of how the capitalist mode of production turns industrial labor into a form of social power, immaterial labor (which is perhaps the most valuable type of human capital today), the productive base of biocapitalism, creates a social space where individuals' identities are heavily influenced by institutions and governmental/corporate entities that stand to profit most from particular configurations of immaterial labor.

Yet as Hardt and Negri's (2004) dialectical understanding of immaterial labor also suggests, such forms of labor that lend themselves to reconfiguration can also be a basis for multitudinal expressions of democratic resistance. That is, as biocapitalism relies on forms of human capital that are highly invested in immaterial labor (deeply important to laboratory work, for example), science education figures prominently in any alternative biopolitical project seeking to produce a different ground for educational life that is not measured by an extractive ethic. Thus, biocitizenship has taken on an added dimension through the biocapitalist expression of scientific literacy being promoted today. This new dimension emerges from the epistemic transfer that takes place through types of immaterial labor privileged by biocapitalism and science education in the neo-*Sputnik* era that works from the understanding that life is a manageable and controllable field. Here, a vital politics has indeed risen to prominence. It is on this axis the recognition that "life itself" has become the target of politics where an alternative vital politics could be located in education.

In setting the concept of scientific literacy within the broader biocapitalist context as I have attempted to do earlier, the field of extractive biopower at work has become more discernable. But what biocapitalism has also done is to create a political context where it is increasingly difficult to bracket off the condition of human and natural life from the activity of science (or for that matter, the activity of learning and teaching science). In other words, in making life one of the most valuable resources to production, biocapitalism has completely dissolved the modern myth of value neutrality that science has turned to time and again for retracting itself from politics. Buoyed by various technoscientific advances that have made it possible to extract value from life, the political stakes of the biocapitalist era thus can largely be seen as now resting firmly on biological grounds—science can no longer claim to solely operate in the realm of nature; it has, in fact, fused the natural and social into complex webs of power in the biocapitalist era. It is also in this space where an alternative scientific literacy could come into being that takes life as seriously as does the biocapitalist productive framework and the form of literacy it seeks to advance. Such a scientific literacy would need to focus on practices that remove life from the promissory and salvationary value framework that underpins biocapitalism. Put differently, a biodemocratic articulation of scientific literacy would need to readjust somatic individuality and sociality in a way that promotes a non-extractive ethic through an alternative ontological zone of production. One possible way to produce such an educational space would be to link learning and teaching science to practices and social movements that are actively resistant to biocapitalist visions of the future, ones that represent cultural practices rooted in communities producing biodemocratic life.

One place to look for an alternative scientific literacy informing a biodemocratic practice is the GrowHaus urban farm and market in Denver, Colorado. Designed and implemented in what one of the co-organizers called a "food desert," the GrowHaus urban farm focuses on local, healthy food production and distribution in a largely minority and working class neighborhood. Taking its cue from the food justice movement that seeks to extend sustainable and ecologically healthy life practices to all communities, the GrowHaus is a strong example of a counter scientific literacy at work that rejects extractive ethics and instead is infused with one that works within and learns from natural and ecological limits.[14] One of the marquee features of the GrowHaus and what makes it unique to other urban gardens within the United States is its integration of an aquaponics system: a sustainable life cycle system that uses "a recirculating process to grow and harvest plants, and farm fish" by using "fish waste [which] works with the beneficial bacteria in gravel and plants, creating a recyclable, concentrated compost" (O'Conner 2010). The GrowHaus represents a space where the science of life is linked to a politics of life in a way that affirms healthy and socially just communities as opposed to treating both as minable resources.

In the GrowHaus example, there are at least two qualities that I suggest should help constitute an alternative scientific literacy that works from a different biopolitical framework. First, much in the same way that Hardt and Negri discuss the use of immaterial labor for creating the skin or ontological ground of a multitudinal expression of democracy, the GrowHaus model reflects a practice of science rooted in a social and political problem—that is—knowledge and technology are utilized in a manner that promotes and respects the healthy limits to life while simultaneously addressing a structural problem within the community: access to healthy and sustainable sources of food.[15] Second, the GrowHaus also exemplifies what Hardt and Negri (2009) see as the often neglected positive aspect of biopolitcs that has the potential for "the creation of new subjectivities that are presented at once as resistance and de-subjectification" (58–59). The GrowHaus, interpreted as a positive biopolitical act, represents a productive zone of life in two senses: natural life in the form of plants, soil, fish, water, solar energy, and bacteria is coupled with a cultural practice of experimenting with alternative political options for growing and distributing food in a community that has been structurally denied access to nonindustrial diets.

As a generative context for producing forms of biodemocratic scientific literacy, urban farms such as GrowHaus could potentially offer students and community members an opportunity to learn and participate in science within an ecosystem and social setting they are actively a part of creating. For example, from the aquaponics system utilized in the GrowHaus, part of a potential pedagogical experience could be based on a model of science that recognizes how human and nonhuman actors are coproducers of science and thus social life. That is, along with their human counterparts, fish, plants, bacteria, and soil are explicit active participants in the production of a life affirming scientific practice: fish create ammonia in their waste that bacteria consume and produce into nitrates that eventually becomes nitrogen, which is used as plant food. In the series of chemical changes that occurs through the interaction between an aquaponics system and planting beds, a perfect example of learning scientific literacy from a situated context emerges. It is one that values and makes clear the connection between biological life processes and community health in a direct, nonabstracted manner. Thus in teaching a lesson on matter and energy, for instance, part of the core curriculum for seventh and eighth grade science standards in most states in the United States, the aquaponics system provides a potentially biodemocratic pedagogical example in this sense: both humans and nonhumans are taken into account in this social setting where science is playing a highly productive role in the health outcome of the community. Here, biopolitics meets scientific practice in a very direct way.

The coproductive act at work here between nature and culture can, therefore, be read as operating with a life-affirming ethic as opposed to an extractive one. Finally, scientific literacy in the GrowHaus model is still very much oriented toward and engaged with life, but in a way that is defined by "resistance and de-subjectification" as opposed to an extractive relationship. In short, the biopolitical circuit is reversed in the GrowHaus model of scientific and community praxis because the relationship between life and forms of biopower has been redefined along qualitatively different power relations.

In turning to biopolitical models such as the GrowHaus, educators, citizens, and scholars could find vibrant examples of scientific literacy that are not invested in treating human and nonhuman life as extractable values. From alternative frameworks such as the GrowHaus and other multitudinal examples, a pragmatic hope can be extrapolated in that the true promise of our social future lies not in learning how to better invest in the circuits of biocapitalism, but instead in learning to create alternative biodemocratic practices that can cultivate autonomous scientific literacies rooted in the needs of communities and planetary health. The future, in other words, largely depends on changing the approach in education from those focused on getting the most *out* of life to ones that make the most *with* life.

CHAPTER 4

Learning about AquAdvantage® Salmon from an ANT: Actor Network Theory and Education in the Postgenomic Era

> Our mission is to play a significant part in the "Blue Revolution" bringing together biological sciences and molecular technology to enable an aquaculture industry capable of large scale, efficient, and environmentally sustainable production of high quality seafood. Increased growth rates, enhanced resistance to disease, better food conversion rates, manageable breeding cycles, and more efficient use of aquatic production systems are all important components of sustainable aquaculture industry of the future.
>
> —AquaBounty Technologies Inc.

> If you talk to the Animals, they will talk with you. And you will know each other. If you do not talk to them, you will not know them. And what you do not know you will fear, what one fears one destroys.
>
> —Chief Dan George

In a recent public meeting held by the Food and Drug Administration (FDA), citizens and political organizations were given a chance to voice their concerns about a "new animal product" application for the first genetically engineered (GE) animal created for human consumption. Not surprisingly, a familiar controversy that first began with Dolly's appearance in 1996 once again found its way into the headlines. What is notable about this recent GE food debate, however, is the growing sense of public helplessness in the face of creatures like the AquAdvantage® Salmon that new types of scientific activity have called into existence. While much of the controversy has centered on the question of whether or not GE salmon should require a food label to disclose its Promethean origin, there has been almost no work that looks at the educational implications of such a remarkable event. What is at stake in AquaBounty Technologies' FDA application (which in all likelihood will be granted), in other words, is not only the question of knowing whether a food item was made by nature or in the laboratory, it also represents an important epistemological crisis of Western modern science that is replicated in the way science is taught and learned; specifically, science education's continued reliance on the modern understanding of science that separates nature from culture which, in turn, leaves individuals and communities unequipped to deal with this question: who and what can be a part of democratic life if the question of life itself has been confused in science's ability to reconfigure nature (wild salmon) through culture (genetic science)?

AquaBounty's claims about their GE fish, ones that are also built into the quickly expanding bioindustries writ large, reflect a politics of science that is the point of departure I take for examining how science education (and education in general) is deeply failing society in what science and technology study theorists have termed the postgenomic era. Postgenomic, according to science and technology theorists Sarah Franklin and Margaret Lock, is a term derived from "the hype surrounding the project to map the human genome [which] has already give[n] way to a new phase of enthusiasm for what has become known as postgenomics. It includes the new science of proteomics, in which an overriding interest in the gene has largely been superseded by interest in complex models of protein and cellular interactions. In connection with developments such as cloning, and in the emergent science of tissue engineering, the cell has reemerged as a central unit of action, temporally and spatially as well as functionally" (Franklin and Lock 2003, 13). What Franklin and Lock as well as other science and technology theorists are pointing to is a massive shift in what is now available to market forces through advances in biotechnological industries such as aquafarming.

Given the postgenomic terrain educators, students, and communities now find themselves, one important problem this chapter takes up is the fact that science education is not adequately preparing citizens for social-ecological settings where the boundaries between "nature," human imitations of nature, and humans have collapsed completely. Given this state of affairs, I am proposing that the way science is being taught and learned must be rethought from an epistemological standpoint that can lead to more democratic relations between humans and nonhumans such as GE food products. Such a move is particularly important given that education in the postgenomic era has failed to recognize and respond to scientific practices that now involve the basic building blocks of life in a whole new productive and consumptive system, one where aquafarming is just one constituent part.[1] At the center of this query, therefore, is a very serious and unexplored pedagogical question that the recent debate on GE salmon and the proposed regulation of how GE food in general will be sold and consumed perfectly reflects the confusing and complex social and political milieu students, teachers, and communities now must negotiate their social lives. It is a social context where what we eat and how we heal ourselves, reproduce, and learn have fundamentally been altered.

The culmination of the human genome project seven years ago, an event that ushered in the postgenomic period, is not just another instance of science uncovering one of nature's most stubborn secrets. The more important triumph that emerged from the genome project was that of the market over one of the last frontiers of non-commodified life: its DNA blueprint. Yet since this landmark achievement in science, one that has turned the basic building blocks of life into usable data to be bought in the form of a patent, sold, and made into exchangeable goods in a rapidly expanding biocapitalist economy, there has been virtually no work done on how such a seismic event affects the role of education in the postgenomic society. Such a lack of any analysis on K-12 schooling and its relation to the mushrooming bioindustries is extremely troubling and arguably represents a form of avoided curriculum. Biocapitalst literacies, what I examined in the previous chapter through the neo-*Sputnik* model of science education, require curricula to boost technoscientific types of human capital. While there is little overt critical analysis of what the shift to a postgenomic arrangement of industries means for our society in terms of education, even just a cursory glance at the discourse emanating from public debate on educational reform today reveals a growing symmetry between the desired school of the future and the biotechnological research lab. Given the current policy and curriculum direction that current education reform is taking, driven in large part by the need for science, technology, engineering, and math (STEM) skills and literacies, how will something like the AquAdvantage® Salmon ever be understood as a democratic problem instead of one that is handled by industry experts, governmental officials, and genetic scientists? In other words, in an educational context where such fundamental questions are decided by experts, how are students to ever sense that they have agency in important social decisions such as the manufacturing and consumption of bioengineered fish?

Developing an alternative scientific literacy appropriate for the postgenomic society, however, is doubly difficult given that standard approaches to teaching and learning science utilize an epistemological framework that separates the cultural world from the natural. That is, one of modern science's distinguishing features is its epistemological allegiance to matters of the natural world and claims of purity from the sociocultural. The work of modern science, in other words, has been understood since modernity as taking place in the objective and knowable universe out there as opposed to the messy human world that is cluttered with things like values, morality, and, above all else, politics.[2] What the debate surrounding wild salmon and its Frankenstein twin AquAdvantage® Salmon reflects is precisely how the Western modern model of science breaks down and is utterly inadequate for building a responsible politics of science and technology within educational settings today. What is needed in its place, and what I outline in the following sections, is an epistemological model for teaching and learning science that is actually capable of capturing the complex series of actors and institutions that are involved in the practice of science in society. To put it another way, scientific literacy needs to be radically rethought in an age where the genes of an Ocean Pout (an eel fish) are spliced with those of a Chinook (king) salmon, implanted in Atlantic salmon eggs, and a corporation patents this process *and* the new species of the fish itself, all while leaving the public's only recourse to understanding such a network of exchanges and relations to the mercy of research done by the leading corporate stakeholder in the aquafarming industry.[3] One of the most important educational challenges that we now face, I contend, is to develop pedagogical responses to such a social arena comprising convoluted sets of relations that can only begin to be understood through the development of a scientific literacy actually capable of critically tracing and making sense of such circuitous routes of "science in action."

To date, there has been virtually no work done on developing a model of scientific literacy that is capable of dealing with the imploding human/nonhuman boundary that has occurred after the human genomic code was deciphered ten years ago. In fact, in looking at the current trends in science education policy and curriculum, the dominant trend is the polar opposite of what should be taking place in terms of preparing students, teachers, and communities with a more appropriate scientific literacy attuned to the postgenomic times in which we now live. As discussed in the previous chapter, perhaps the best example of the dominant trend in scientific literacy is the one promoted by the National Academy of Sciences' *Rising above the Gathering Storm: Energizing and Employing America for a Brighter Economic Future* (RAGS). In the most comprehensive science education policy recommendation publication since the *Sputnik* era's National Defense and Education Act, the type of scientific literacy being called for in this highly influential document is directly related to the burgeoning biomedical and bioscience industries' need for a steady supply of capable knowledge workers. As a strategic plan, *RAGS* targets curriculum design, teacher training programs, the recruitment of a greater number of undergraduates into the STEM fields through financial incentives attached to teaching service, and the deregulation of international visa policy for science and engineering students from other nations across the globe. The movement to increase bioscientific production through education represents (which is also backed by the Obama administrations' signature education legislation Race to the Top) a push to shape scientific literacy as a tool for powerful industries and global economic competition. Instead of the development of a capable literacy for citizens who find themselves in a maze of ecological and social problems without a map, neo-*Sputnik* models of scientific literacy only make the problem more confused.[4]

Alternative pedagogical strategies in the field of science education have begun to draw connections between science and society in meaningful ways. For example, the "teaching the controversy" method emphasizes how controversy, such debate about genetic cloning in scientific communities, can be a productive point of departure for science teaching and learning as it forefronts how science is involved in cultural questions regarding the limits to science (Hines 2001).[5] Other science educators and theorists have looked at ways to integrate science and technology studies (STS) approaches and socioscientific issues (SI) into science education contexts in an effort to broaden the meaning and scope of scientific literacy (Aikenhead 1994, 2006; Zeidler et al. 2005; Linder et al. 2011). As Erminia Pedretti and Joanne Nazir put it in their survey of the history of STS work in science education, "at the macro level, STSE education situates science in rich and complex tapestry—drawing from politics, history, ethics and philosophy. It presents an opportunity to learn, view, and analyze science in a broader context, while recognizing the diversity of needs of students and classrooms...STSE, in its many forms and currents, brings relevancy, interest, and real-world connections to the science classroom" (2011, 618).[6] While the articulation of scientific literacy in the work of these theorists and educators is productive and much needed, none utilize actor network theory (ANT) to approach the problem of how to untangle the networks of actors that do not neatly stay within the confines of either nature (the domain of science) or culture (the domain of humans and their politics) in postgenomic society—A task I take up in the following through the eyes of salmon and the path of an ANT.

This chapter applies ANT to the life of salmon to demonstrate its pedagogical uses for developing an alternative scientific literacy in both science education and social studies contexts. The first frame of the actor network map of salmon comes through an engagement with the emerging field of environmental history. Here, I argue that part of preparing students and citizens for postgenomic politics is to cultivate actor network understandings of the historical relations between humans and nonhumans (the natural world in modern terms) in the industrial and colonial model of development. More specifically, part of developing a more appropriate scientific literacy for the postgenomic era is to understand how science and technology have shaped our understandings of human and nonhuman relations in a way that blinds us to alternatives that already existed (and that are still practiced today) on this continent before colonial expansion and the triumph of the extractive industries, such as timber, mineral, water management, and commercial fishing. The second line of investigation I use to flesh out ANT's pedagogical potential is an analysis of how our understanding of salmon has been reconfigured through technoscientific practices and discourses that treat nonhumans as discernable and researchable entities of nature and thus something only to be understood through science. In this section, I make the claim that in order for fish, humans, and genes to be involved in a democratic relationship, the way science constructs a nonhuman such as the AquAdvantage® Salmon, which is both a GE food product and a public entity, needs to be critically reinterpreted through ANT. In so doing, I argue that what is needed is a model of scientific literacy that can highlight and trace technoscientific practices, skills, and literacies in society that have been absorbed into the National Academies' latest science education standards framework. The final frame I develop brings together the previous two and adds a third to construct a pedagogical example of how ANT can be applied in educational settings to help cultivate a viable political framework for education in postgenomic society based on an ethic of the common. Here, I build upon my own work in the seventh and eighth grade science education settings by using the actor of salmon to help demonstrate the practice of creating actor network maps, a pedagogical practice that works toward rethinking the foundations of democratic education from the standpoint of human and nonhuman relations.

Before I proceed into my construction of frames for rethinking the pedagogical role of salmon as an example of how to think like an ANT, it first needs to be made clear exactly what an ANT is and why it provides a useful perspective for the question under investigation in this essay; namely, how does education respond to a political terrain where a human-made fish swims between nature and culture but is only treated as an object to be neatly fitted into its growing pen enclosure and the fish counter at the supermarket?

The World of ANTs

The objectification of nature taught through the epistemological mode of modern science, as Bruno Latour (1993) has argued throughout his work to develop ANT, perpetuates a relation with the natural world that denies the sociality of objects and their role in the discourse and practice of science:

> Yet the human, as we now understand, cannot be grasped and saved unless that other part of itself, the share of things, is restored to it. So long as humanism is constructed through contrast with the object that has been abandoned to epistemology, neither the human nor the nonhuman can be understood. (136)

As Latour has deftly pointed out, the object of modern science's gaze has ceased to be a silent feature in the social and political terrain. According to him, "networks" of science and technology create both "human" and "nonhuman" agents in a collective that is mediated through political discourse and practice, on the one hand, and scientific discourse and practice, on the other, with each protected through its exclusion of the other.

Shaping Latour's work is what has come to be known as ANT, a perspective codeveloped with fellow science and technology theorists John Law and Michel Callon. For them, "the origin of this approach can be found in the need for a new social theory adjusted to science and technology studies" (Latour 2005, 10). One of the central contentions driving ANT for Latour, Law, and Callon, thus, is the need to develop a social theory that can trace and give more accurate accounts of the associations between objects, humans, and political arrangements:

> ANT claims that we should simply not believe the questions of the connections among heterogeneous actors to be closed, that what is usually meant by "social" has probably to do with the reassembling of new types of actors. ANT states that if we wish to be a bit more realistic about social ties than "reasonable" sociologists, then we have to accept that the continuity of any course of action will rarely consist of human-to-human connections (for which the basic social skills would be enough anyway) or of object-object connections, but will probably zigzag from one to the other. (Latour 2005, 75)

The formulation of a sociology of science and technology, for Latour, requires an epistemological rethinking of how both political philosophy and philosophy of science have traditionally treated the problems of reality, certainty, truth, and political agency. As the nexus of focus for these two modern fields, each discourse has seemingly separately perpetuated modes of enquiry and political organizations that obscure the reality that they actually codetermine each other.

The most important aspect of Latour's work that I want to emphasize here is his examination of the epistemological disconnect that has its origins in two disciplines of knowledge production in society: sociopolitical theory and scientific theory. The sociopolitical concentrates on the affairs of the state and its citizens, while science deals exclusively with the objects in its self-designated domain of nature—thus, both realms miss the articulation and mutual influence of science, objects, and the natural world. This is another important insight that Latour develops in his critique of modern science: that the dual dichotomies erected between human and nonhuman culture produce a false separation between nature and object and human activity. The first dichotomy is constituted through the work of "purification," which "creates two entirely distinct ontological zones: that of human beings on the one hand; and that of nonhumans on the other." The second dichotomy is constructed through "translation" (or mediation), which "creates mixtures between entirely new types of beings, hybrids of nature and culture" (Latour 1993, 10). The concepts of translation and purification are important to draw out here, as they reveal how humans and science coproduce social experiences and specifically, how current notions of scientific literacy fail to equip individuals with an adequate tool for political life in the postgenomic era.

Translation and purification represent for Latour how, epistemologically, modern science simultaneously adopts an ontological distinction between objects and humans while proliferating hybrid network relations made up of culture, nature, technology, and the specialized discourse that is produced through these encounters. Purification is the practice of sharply distinguishing objects in the world, asocial in nature, from the observer who is situated within the social and cultural world. The practice of purification also takes place above the horizontal line that separates it from translation (or mediation). In the zone beneath, the work of translation is exhibited through hybrid networks constituted of both human and nonhuman actors; these are involved in a process of mutual agency that codetermines social environments and ways of communicating through networks of standardized practices. The laboratory, as a producer of hybrid networks, exemplifies this exchange between humans and nonhumans: "We live in communities whose social bond comes from objects fabricated in laboratories; ideas have been replaced by practices, apodeictic reasoning by controlled doxa, and universal agreement by groups of colleagues" (Latour 1993, 21). The contradictions programmed into this process perpetuate an antidemocratic relation between science and society that stems from the modern epistemological framework of science: we approach science with a lens that was forged through the understanding of the natural world as an other, but we can only do science in society if the other becomes a part of the culture of science. Once nature or the object has become a part of science, it gains agency and determinative qualities in culture—that is, objects are not voiceless since they help determine how science explains and interprets the world.

So long as this aspect of modern science is misrecognized, science education and, by extension, scientific literacy will continue to utilize an antidemocratic epistemological modality.

In the following sections, I attempt to untangle how purification and translation operate as separate activities in the way science is taught and learned by applying actor network to the nonhumans of Pacific salmon (Oncorhynchus) and its GE offspring, the AquAdvantage® Salmon. The lines of flight that comprise the actor network of the Pacific salmon and AquAdvantage® Salmon that I develop in the following sections by no means exhaust the entirety of actors involved in the story of the salmon and the humans it convenes with. Rather, I am utilizing three frameworks that bridge the cultural and natural or rather disrupt the act of purification and amplify translation to tell the story of Oncorhynchus and its GE relative as a way to model how ANT can be used in educational settings and specifically for developing a more adequate scientific literacy for the postgenomic era. As the most well-known anadromous fish, one that begins and ends its life in fresh water but grows to maturity in the ocean, Pacific salmon are truly a unique actor in which to chart an actor network.[7] Much in the same way that they can pass between a vast saline world to a multitude of urban and rural fresh water veins and arteries, Pacific salmon perfectly embody network life. The ecological devastation of deforestation practices and suburban overdevelopment that destroy stream and river spawning beds, the threat of genetically engineered hybrid species to wild salmon gene pools, and the catastrophic effects of aquafarming on young wild salmon fry that are a result of sea lice explosions in migratory salmon zones, all represent a context in which the human and nonhuman have become inextricably a part of one another's lives. It is this world that we must dive into and see how separating the natural from the cultural is not only an impossible feat today, but also one that makes very little sense when looking through the eyes of salmon.

First Tributary: Understanding Human and Nonhuman Relations through Environmental History

The field of environmental history is a relatively young one.[8] Alfred Crosby's *The Columbian Exchange: Biological and Cultural Consequences of 1492* (1973) and his later *Ecological Imperialism* (1986) are foundational texts that mark the inception of the field. More recently, however, William Cronon has pushed environmental history in new and highly productive directions. In his seminal *Nature's Metropolis: Chicago and the Great West* (1991), Cronon reveals in intricate detail the transformation of the upper Midwest prairie ecosystem into the greatest industrial city on the edge of the vast Western frontier of the United States. Here Cronon's focus, as opposed to traditional approaches taken in histories of place that center on great figures, momentous events, or defining battles, rests on the interaction between humans and the land that both fed city growth and the machines that helped bring the nineteenth century metropolis into being. What unfolds in Cronon's history of Chicago and the great prairie grass lands of the West is an expansive network of interactions between industry, railroads, canal systems, meat packing and grazing, lumber extraction, city building, and native removal and white settling that tells the story of how such an assemblage produced a great upheaval and violence to the land and its original inhabitants. Yet at the heart of Cronon's (1991) analysis is a highly original method for interpreting both history and the co-constructive role humans and nonhumans play in its making:

> The urban-rural, human-natural dichotomy blinds us to the deeper unity beneath our own divided perceptions. If we concentrate our attention solely on the city, seeing in it the ultimate symbol of "man's" conquest of "nature," we miss the extent to which the city's inhabitants continue to rely as much on the nonhuman world as they do on each other. We loose sight of the men and women whose many lives and relationships—in the city or the country, in factory of field, in workshop or countinghouse—cannot express themselves in so simple an image as singular man conquering singular nature. (18)

The implications of such a reworking of historical interpretation or rather an opening up to the nonhuman as an historical actor are indeed a radical shift from traditional historical narratives of the West. However, it is perhaps in Cronon's controversial essay a few years after *Nature's Metropolis* that such a method of rereading history from the vantage point of human/nonhuman relations came into full bloom. In his essay "The Trouble with Wilderness; or, Getting Back to the Wrong Nature," Cronon (1995) expands his thesis of viewing history as an assemblage of human values, natural objects and features, and industry by arguing for the jettisoning of the sacrosanct idea of wilderness that seems to be so deeply entwined with the myth of American progress and the ethic of human domination over nature that fuels it. For Cronon, the idea of wilderness in US culture and history has largely represented a place of redemption and renewal. It is the place understood to be beyond the city, town, or frontier where one can find and start a new life, stake claim to "uninhabited land," or seek religious solace in the sublime glow of nature (Cronon 1995). In short, wilderness according to Cronon has long held a transcendental place in the cultural imagination of the United States. However, such a view of wilderness also allows for the normalization of cultural perspectives that omit the fact of how "the removal of Indians to create an 'uninhabited wilderness'—uninhabited as never before in the human history of the place—reminds us just how invented, just how constructed, the American wilderness really is . . . Indeed, one of the most striking proofs of the cultural invention of wilderness is its thoroughgoing erasure of the history from which it sprang" (Cronon 1995, 79).

In troubling normative understandings of nature such as the concept of wilderness, Cronon and the field of environmental history generally open up an entirely new way of understanding the relationship between the human and nonhuman and particularly the ways in which both are involved in constructing social and cultural spaces. From the perspective of the salmon, the dissolution of the dichotomies between wilderness/civilization and human/nonhuman that the field of environmental history has put at the center of its project becomes even clearer. The history of the fish itself is a history of a false wilderness ethic that assumes an unending supply of silver-finned creatures while also perpetuating a blindness to the way an industrial land ethic impacts fish as well as the management practices of the peoples who originally lived with salmon in a sustainable relationship for eons. Pacific salmon, in other words, are deeply representative of a pathology that still very much lives today in the way individuals, corporations, governmental policies, and educational curriculum (especially science education) regard natural resources. One of the consequences of such a view of nature as outside cultural and social settings is an epistemological perspective that privileges scientific and technocratic solutions to crises that are a product of this very same bifurcating model. The use of fire to fight fire, in other words, has been the reaction to the decline of the Pacific salmon as well as other untold casualties of the nonhuman world.

As environmental historian Joseph Taylor has pointed out, one of the best ways to understand the plight of the Pacific salmon is to compare the industrial and technoscientific approach to fisheries that white colonization brought West to that of the aboriginal fisheries that existed for centuries prior in the Oregon country. However, this look back, as Taylor warns, is not a return to a mythical Eden, a place where American Indians are romanticized as a monolithic group who lived in perfect harmony and balance with nature as depicted in many modern literary accounts, media and cultural representations (the crying American Indian image from the 1970's anti-littering public service announcement), or even iconic images used to sell "wild" places to white settlers and tourists.[9] What Taylor (1999) does suggest is that "the aboriginal salmon fishery provides a useful lens for analyzing the intersection of economy, culture, and nature in the fisheries because Indians did influence salmon populations" (13). Moreover, it is particularly important to look at the history of the environmental relations between salmon and the diverse indigenous peoples of the Oregon country because "the scale of the aboriginal fishery and the Indians' dependence on salmon posed a significant threat to runs, yet Oregon's rivers still teemed with salmon while whites arrived. How did this happen?" (Taylor 1999, 14).[10] What needs to be understood, in other words, and what environmental histories such as Taylor's can help elucidate, is an alternative model of management that once made up (and that are still practiced) the network of relations between humans and nonhumans in the world of salmon.[11] Such models of land and water management offer a real and practical guide to a sustainable scientific literacy, one that could come from what indigenous scholars such as Winona LaDuke (2002b), Gregory Cajete (1999), and Oscar Kawagley have called traditional ecological knowledge (TEK) or ethnoscience (Kawagley et al. 1998).[12]

Environmental history thus also helps point to existing alternative technologies and land ethics that can be looked to for developing alternative scientific literacies. For instance, the different technologies American Indian groups use (and used on a wider scale precontact) to harvest fish such as seines (nets suspended by floats and weighted at the bottom usually in shallow, calmer water of a river or an estuary), dipnetting at sections of water where falls or rocks form natural fishing holes, or the weir technique that consists of a fence and lattice system that creates a permeable gate across the river are all fishing methods and tools that point to more sustainable and healthy relations with nonhumans. Given the nonindustrial design and ethic associated with these fishing technologies, many American Indian groups had great success (and still do with gill nets, e.g., among other methods) with such tools while not depleting salmon numbers in any significant way (Boxberger 1989, Taylor 1999, LaDuke 2002b). Again buried by technocratic approaches associated with the ethic to dominate and extract value from nature that whites brought West, as Taylor's environmental history of salmon draws out, are real and existing fishing technologies that are more in balance and scale with their local ecology.

What Taylor's environmental history of the Pacific salmon brings to the foreground is thus a tension between two cultural frameworks for understanding and interacting with nonhumans. It is on this point that environmental history provides an indispensible place from which to link ANT research and the development of a more appropriate scientific literacy. The reason environmental history can serve this function has to do with the fact that one of its primary strengths is the tracing of relational histories between humans and nonhumans (both natural and human made). The creation of a pedagogical bridge between traditional Western approaches to science education and alternative epistemologies (and ontologies) with nature, such as the one Taylor points to in his history of the Pacific salmon crisis, is a particularly important project since nature is positioned in these histories as an actor and not an object to be defined and understood solely through science. In this sense, from an environmental history lens, two cultural systems of land and water management are cast into relief and lend themselves to comparative analysis, which also opens to a new way of thinking about nonhumans that modern science alone rejects. Also, in creating such a juxtaposition, what also becomes apparent are different ethical groundings for a politics of nature. From a comparative standpoint, germs and technologies of control and management that accompanied the increasing Anglicization of the Oregon country are highlighted as key actors in the environmental history of the Pacific salmon's riparian biome—ones that cleared the way for a technocratic and exploitative ethic to establish relations between humans and nonhumans in the conquering of the West by whites. That is, smallpox, measles, dams, land and water treaties, legal treaties and statutes, and tuberculosis could arguably be seen as the most important actors in the conquest of Western lands by European peoples, and thus must be taken into account when considering the history of salmon and the other nonhumans involved in their fate (Crosby 2003, 1986 Headrick 1981; Wilkinson 2000; Mann 2006).

From the opened field of vision, environmental history brings into view a vast array of actors that have helped populate the history of human contact with Pacific salmon, constituting a more accurate picture of its political terrain. Thus in examining how "dependence on salmon yielded forms of respect for the fish that sustained life, and respect shaped human actions that retarded consumption" and how "Indian culture and economy produced a sustainable tension between society and nature" a political framework in which humans (indigenous peoples of the Oregon country) and nonhumans (salmon) exist in a reciprocal and co-constructive manner. This framework opens up a pedagogical space in which to rethink scientific literacy from the standpoint of the nonhuman (Taylor 1999, 14). Herein lies the strength of ANT working through the lens of environmental history: in focusing on the relations and associations between land, water, fish, technology, knowledge systems, and different cultural ethics with nonhumans, an epistemic landscape is opened up that is both more accurate and more educative for developing a viable and sustainable politics that reflects the way science and technology have been blurring the boundaries between culture and nature, human and nonhuman for some time now.

Second Tributary: The Scientific Making of Salmon

As our salmon has moved upstream a bit further, making its way through the currents and sand bars made visible through the perspective of environmental history, it is now time to follow the fish into the laboratory, courtroom, and classroom. If the previous section's aim was to show how environmental history can be a useful tool for developing a scientific literacy informed by ANT, then here the focus is on the anatomy of salmon as it is constructed through AquaBounty's FDA application that lays out its "product definition" for the GE animal it is seeking to patent and sell. Put another way, here the focus is not on the past life of salmon, but rather its public life articulated through the only acceptable mouthpiece for nature: scientific experts. Yet, as I argue in this section, what can also be seen in this litigious and scientific stretch of the river is a distinct model of scientific literacy that science education is increasingly embodying in the high-stakes, globally competitive bioscientific and biotech-driven economy. In looking at the new science education standards just released by the National Academies, a symmetry between the type of science it takes to produce the AquAdantage® Salmon and the new standards framework is remarkable. In the National Academies' conceptual streamlining of STEM education, a central focus of the recent science education standards overhaul, laying the groundwork for biotechnological and bioscientific scientific literacy is clearly one of its primary goals. The connection that is evident, and what this section investigates, is how the mode of scientific literacy needed to produce the AquAdvantage® Salmon or any new commodity in the mushrooming biofarming industries requires a particular type of scientific literacy to be produced from the education system. In particular, a scientific literacy that is geared toward manipulating and commodifying forms of life, that also produces a public understanding of science largely detached from scientific and legal decisions, and that releases biotechnological forms of life into public life. It is this reflexive relationship between the scientific literacy that underpins nonhumans such as the AquAdvantage® Salmon, and science education that must be focused on while keeping one of ANT's strongest qualities in view: an understanding of how science involves nature in its social activity yet denies it any status other than nature, a decidedly apolitical designation.

As Taylor's environmental history of the crisis in Pacific salmon fisheries points out, science has been intervening in the life of salmon since the mid-nineteenth century in the United States. From the context of the Atlantic salmon fisheries crisis on the East coast, the science of fish reproduction in hatcheries was developed in the nineteenth century as a rational response to the problems of overfishing and destruction of fish habitat from processes of industrialization. Many species of fish that once flourished in streams, rivers, lakes, and the Atlantic Ocean, it could be argued, were both a symptom and a pathology of the industrialization of the United States: a symptom in that their life became entangled with industrial relations and a pathology because their ultimate fate rested in scientific and technological responses to increasingly eviscerated "natural" conditions for life. One of the most influential ways fish culture spread as a scientific panacea to mounting fisheries crises across the United States in the late nineteenth and twentieth centuries was through institutional bodies such as the US Fish Commission. Through such agencies, "fish culture became the preferred tool of management because it offered to produce an endless supply of fish. Salmon hatcheries seemed to facilitate economic progress while alleviating resource conflicts. It was no wonder that Americans embraced fish culture, but neither were their choices timeless. Hatcheries were a logical product of a context. The logic was shaped by particular perspectives on government, science, and nature" (Taylor 1999, 98). The advent of fish culture as a way to manage a problem that comprised a much more complex set of relations than the simple equation of too few fish for industrial consumptive patterns + technical/scientific fix = solution (more fish) represents the dominant societal approach that continues to this day with other ecological and social problems (i.e., carbon sequestration, clean coal, bioengineered food, etc.) that the GE salmon debate also perfectly represents.

What is important to recognize in this analysis is how such a scientific literacy carries with it a certain epistemological practice that normalizes students, teachers, and citizens to the ways in which nature continues to be perceived as a fixable and extractable entity as opposed to a network of actors already involved in the productive act of science and, by extension, our social realities. In other words, what needs to be made clear is how hatchery fish or AquAdvantage® Salmon are part of a larger model of scientific literacy that fails to take into account how wild salmon (and GE salmon for that matter) can also teach us something other than a market-based urge to create more biological entities that resemble salmon. Rather, salmon can teach us that nonhumans and the network of actors they are associated with (i.e., dams, fishing industry, fish biologists, management policies, indigenous rights, etc.) need to be part of a viable politics of the future so that more actors, besides AquaBounty investors, bioengineers, and FDA policy makers, can be involved in making decisions about future relations between humans and nonhuman actors such as salmon. I return to this democratic question in the last section. But first it will be helpful to examine the connection between the science of GE fish and the latest iteration of content standards for K-12 science education.

The FDA hearing that took place in September 2010 had as its purpose to review the "risked based, scientific process for determining the safety and effectiveness of the AquAdvantage® Salmon produced by AquaBounty Technologies Inc." (Food and Drug Administration Center for Veterinary Medicine [FDACVM] 2010). The life of our second salmon, the AquAdvantage® Salmon is one that is very much alive in this public performance of science, and, as such, the FDA report is a vivid example of how science is involved in much more than laboratory work. The aim here, therefore, is to determine how the productive partnership between the biotechnological food industries, of which aquafarming is the fastest growing, and nature have created a fish that symbolizes both the future of the dinner table and the science education classroom.

The increasingly symmetrical relationship between the laboratory and school should not come as a surprise given that the AquAdvantage® Salmon swims in a river that has gone in volume from "1 million tonnes in the early 1950's" to "51.7 million tonnes and a value of $78.8 billion" in 2006 (AquaBounty Technologies 2010). In the wake of such a dramatic growth trend, AquaBounty's 200-page document under review by the FDA for a "new animal drug application" lays out in great detail the science behind its new finned invention as well as the powerful industry interests that drive it. Moreover, in this public construction of the AquAdvantage® Salmon, what is apparent is a model of scientific literacy that accurately represents the co-constructive act of blending science and nature, human and nonhuman, a relationship that biotechnological industries are built upon. What it also discloses is how a hybrid entity is produced while simultaneously reinforcing the view that nature, even in a reconfigured form such as the AquAdvantage® Salmon, is still only explainable through the act of purification, or making the fish unequivocally a thing of nature and thus only accountable to the scrutiny of scientific experts, auqafarming investors, and pro-market governmental policies.

In drawing out the type of scientific literacy operating in AquaBounty's FDA report, one of the most productive places to look is at one of the key claims the company and its scientists are attempting to make: that the AquAdvantage® Salmon is *almost* identical to wild Atlantic salmon with the exception of the rDNA construct that AquaBounty scientists created by combining Pacific salmon (the Chinook's growth hormone gene) and Ocean pout (for its "transcriptional regulatory elements") gene segments.[13] In a way, this problem for the aquafarming industry, its need to explain how its animal products are similar enough to their natural counterpart so that they can be safely consumed and raised, is one that directly reflects what biotechnological literacies produce as a form of scientific practice: making objects that are seemingly recognizable and assessable through the lens of technoscience and by the experts who wield its power.

Thus, what we see within the context of aquafarming's productive framework is a mode of scientific literacy that is both created and judged by its own criteria such as the "molecular characterization of rDNA construct." That is, the only way to evaluate the AquAdvantage® Salmon is through scientific discourses and practices that are self-referential to the very productive apparatus from which they came. In working backward, however, from the scientific standards mandated by the FDA, specifically evidence of the "molecular characterization" and "phenotypic characterization" of the new animal product, what emerges is a scientific literacy driven by imperatives that seek to control, manage, and market life at the molecular level. The following criteria established by the FDA for defining a "new animal product" represent the terrain in which such a scientific literacy is working and demark the level to which nonhuman and human life are now involved with one another in postgenomic configurations of nature/culture: a description of the source(s) of the various functional components of the construct (the unique rDNA sequence created by scientists), the sequence of the rDNA construct, the purpose of the modification, details of how the rDNA construct was assembled, the intended function(s) of the introduced DNA, and the purity of the preparation containing the rDNA construct prior to introduction into recipient animals or cells (FDACVM 2010).

Embedded in the scientific criteria presented by the FDA for evaluating any "new animal product" is the hidden story of how technoscientific literacy functions. This type of scientific literacy can be distilled into at least three distinctive processes: (1) creating the ability to extract from "nature" genetic sequencing from two different species of fish (Chinook salmon and Ocean pout) and (2) a recombination of this genetic information into an rDNA construct that can be injected into a third "natural" object, an Atlantic salmon egg, while (3) producing an entirely new species of which proprietary rights are one of the primary objectives. At work here is a distinctly technoscientific model of science, what science and technology study theorists have identified as the foundation to laboratory and scientific work in the postgenomic era.

The term technoscience designates what Bruno Latour and Donna Haraway among other science studies theorists have argued in their work is the conflation of knowledge production (formerly science) and technologies (formerly applied science) into a more fluid, codevelopmental model (Haraway 1997, 2008; Latour 1987, 1993, 2005; MacKenzie and Wajcman 1999). Put another way, contemporary science cannot be understood outside of the powerful technological tools that help bring "discoveries" such as GE animal products into existence. For example, without a computer and complex software programs, the mapping of the human genome could not have taken place. More precisely, it is such a coconstructive relationship between technology and science that science studies theorists point to as the synergy underlying knowledge-producing sites such as university/corporate biotech research partnerships, genetic start-up labs, or biopharmaceutical giants such as Merck. Calculating and sorting endless series of gene data, in other words, are an impossible task without supercomputers and machines capable of identifying and separating microscopic genetic material, *and* the human beings trained to make sense of such data. It is the demands of technoscientific forms of production, such as those created by AquaBounty Technologies, that scientific literacy in our postgenomic society must be understood as coevolving with the largest and fastest growing sectors of the economy, such fields as biofarming, biomedicine, and the biosciences in general.

What technoscientific types of production require, as can be deduced from AquaBounty's FDA report, are scientific workers who are well educated in not only traditional scientific methods but also in bioengineering, biochemical science, and mathematical modeling. Furthermore, what we see when we look at the case of the AquAdvantage® Salmon is a model of science that is not on the horizon, but a type that is already present and quickly becoming the dominant paradigm informing how science is taught and learned. In the example of the FDA report that perfectly embodies how technoscience melds science, engineering, and mathematics in a prized commodity of the aquaculture industry, we can also look into the future of science education. The science education content framework recently released by the National Research Counsel of the National Academies has unmistakably seen this future and has responded accordingly. Here, one of the larger goals of the new standards framework is to create a way of teaching and learning where "engineering and technology are featured alongside the natural sciences in recognition of the importance of understanding the designed world and of the need to better integrate the teaching and learning of science, technology, engineering and mathematics" (National Academies of Sciences [NAS] 2010, 8). It is not a coincidence that the complete reworking of the national science education standards has taken the approach of "crosscutting" the natural and life sciences, mathematics, and engineering as a unifying theme.

In mixing the knowledge systems and application procedures of science, mathematics, and engineering, the National Academies' framework is responding to a social need (knowledge workers trained in technoscientific practices) while also perpetuating the notion that scientific and technocratic practices are the principal fertilizer of our economic and social futures. That is, the new science education standards cannot fully be understood outside of the political context from which they were born, one where the educational system in the United States is increasingly being viewed as seriously deficient in areas of scientific and mathematical aptitude, which, in turn, signals an emergency in the knowledge pipeline to the fastest growing industrial sectors currently driving the future economy of the United States, bioscientific and biotechnological production. Indeed, the National Academies' new framework for K-12 science standards, in many ways, is responding to the economic and by extension national security need that hinges on a strong and capable technoscientific workforce: "It is important to note that though we make little explicit reference in the framework to biotechnology and medical applications of science (except that bioengineering is included under the general definition of engineering used here) we do not thereby intend to downplay their importance. In fact, much of what is said about engineering could be applied in that context as well, if you think of bioengineered organisms, medical treatments, or drugs as something that you design, and that inclusion is intended here" (NAS 2010, 12).

What I am suggesting is both the implicit and explicit goal of the NAS's standards framework, is a model of scientific literacy that is attuned to technoscientific modes of labor *and* a corollary public understanding of science. On the one hand, what is evident in the NAS's new science education framework is the move to crosscut disciplines within the sciences. Whereas in the past science and mathematics education has largely been theorized and practiced as separate entities, the new standards seek to coalesce science, math, engineering, and technology into a streamlined educational approach. Part of the reason for merging three formerly distinct educational areas is, as I have pointed out earlier, because the technoscientific workplace has already conflated these subjects and practices going on for many years now in genetic and pharmaceutical laboratories across the globe. In this sense, the NAS's science education framework is simply mirroring what is already going on in the real world, albeit this real world is also one that is heavily driven by high investments into biotechnological and biomedical sectors of the economy—the biggest single driver of the global economic future (Rose 2007). On the other hand, another effect of crosscutting science education is a naturalization of "bioengineered organisms, medical treatments, or drugs" that technoscientific modes of research and production allow humans to create. In other words, at what point in the construction of a scientific literacy that is based on and promotes technoscientific skills are students, citizens, and educators able to question what organisms, medicines, treatments, or drugs are created and for whom? Or, more in line with our salmon, how does the public make sense of the AquAdvantage® Salmon when the science that bore this fish is already assumed to be the only worthy judge of its own merits?

From the perspective of ANT, what the new framework for science education standards advances is an ever-deeper epistemological blindness to purification (science is only involved with the objective world) while also simultaneously striving to produce more educational subjects who are involved in the creation of objects of translation such as the AquAdvantage® Salmon. In addition to normalizing technoscientific literacies, however, the NAS's new standards framework also has an important political dimension that ANT helps us ascertain. If we turn to our salmon as an example of a politics of science that embodies the act of purification, the implications for education immediately become clear. That is, when we look at the AquAdvantage® Salmon from the standpoint of ANT, the question arises as to how the modern concepts of nature and culture are preserved as separate domains through technoscientific literacies? From a modern scientific perspective, this question is a conundrum, a query that makes little sense given the fact that science not only communes with the natural world but also brings things back into the cultural world and sets them loose on a public that is not equipped to deal with such escapees. It is this very act being conducted in the FDA hearing on the AquAdvantage® Salmon that the NAS's new science education standards framework also promotes: the proliferation of scientific discourses and practices designed to bear hybrid entities while purifying their existence in society as safe and consumable goods.

By following our salmon through the laboratory, courtroom, and classroom, what emerges is a type of scientific literacy that is severely hobbling the ability of students, teachers, and communities to enter into democratic relations with nonhumans such as the AquAdvantage® Salmon and its all too human creator, AquaBounty Technologies. Yet what is also taking place at the epistemological level of teaching and learning is the normalization of technoscientific modes of science education that reflects more of what is going on in the high-stakes types of research and development of start-up genetic labs than it does with the realities of social and ecological risk that make up the peoples' experiences in everyday life. What educators, students, and citizens are faced with is indeed a double bind. First, where should educators start to look for an alternative model of scientific literacy that can critically engage with the dominant technoscientific literacy driving education in the postgenomic era (one driven by powerful industry and governmental interests), and second, one that provides epistemological and democratic tools for brining the nonhuman and human worlds together into a political collective where GE fish and their fate are determined by more than just expert fiat? In the final section, I will explore such a possibility, namely, how science education can be reconfigured to meet the demands brought onto social life by the continued presence and participation of the nonhuman with the human world. Since such a reconstructive task is an enormous one, I only focus here on how scientific literacy in educational settings might begin to alter the relationship between nonhuman and human political life. Here again we are going to draw upon ANT to help constitute what such a literacy of science should entail. All we need to do is follow our salmon out of the tributaries and into an estuary system that hopefully can lead to an ocean of greater commonality.

Estuary of the Common: Learning from Pedagogical Nonhumans

One way to go about exploring how the path of an ANT can help develop a well-tuned scientific literacy, one that can lead to an epistemological untangling of the work of purification and translation that technoscientific practices are built upon, is to look at what such a shift in educational approach might entail. In work that I have conducted with seventh and eighth grade science students, teachers, and community members to develop a model of scientific literacy informed by ANT, community problems and controversies were used as the basis of enquiry for constructing actor network maps.[14] For example, one actor network research project took a recent local oil spill that contaminated multiple socio-ecological (both urban and rural) zones to frame the process of enquiry for student actor network maps. Student actor network maps traced a constellation of actors that modern scientific approaches treat as separate symptoms of the catastrophe (such as water testing in the creek, park pond, and rivers most affected; identifying the cause of the leak; etc.) at best while leaving other important actors out of the picture entirely. Unlike traditional science education approaches, students researched key actors in their project that more accurately depicted the network of relations involved in oil extraction and processing. These actors included one of the largest petroleum corporations in the world (Chevron), cutthroat trout, aquatic insect extinction which triggered trout extinction, an 800-mile oil pipeline, the US Department of Transportation (the governing body for a cross state pipeline), waterfowl, a riparian biome, oil refineries, oil drills, tankers, semis, and university scientists studying the effects of the spill, just to name some of the most influential. The production of actor network maps also integrated core standards for seventh and eighth grade science students, such as chemical properties, physical properties, food web, energy flow, producer, consumer, environment, fossils, energy, potential energy, and geology. In this way, actor network maps put these concepts into motion by showing the fluid qualities of actor networks, giving students an opportunity to expand scientific literacy in ways that more precisely reflect science and its relation to socio-ecological settings while also practicing sound scientific skills such as observation, the collection and evaluation of facts, and creating and revising research questions. Students also participated in the presentation of their research to their community, modeling the practice of disseminating and sharing knowledge with a larger group of stakeholders in the community.

Building off these practical experiences of implementing ANT as a pedagogical tool, the following sections provide some preliminary curriculum design features that offer a possible outline for extending ANT's educational uses. Part of what this means, however, is the release of some hard-held beliefs and entrenched pedagogical approaches to learning and teaching science in society. Again, we must turn to the fish and ANT to think through this difficult proposition.

Transitional Frames to ANT Literacies

Part of the reason salmon make for such a good example of an actor to demonstrate the pedagogical potential of ANT is, in terms of nonhumans, salmon embody network life. They literally have been swimming in and out of nature and culture for millennia; their highly migratory lives, that is, take them from desolate rivers and streams to urban waterways, heavily managed rivers, and to the seemingly endless openness of the ocean. Salmon have also been embroiled in human controversy since at least the mid-nineteenth century, ranging from indigenous fishing rights and the take over and management of fisheries by whites, the creation of fish hatcheries, to concerns over rapid population decline along with its ultimate hubristic response: the AquAdvantage® hybrid salmon. Given this complex set of relations, salmon make for a very rich area of enquiry that lends itself perfectly to an educational approach for teaching and learning science from an ANT perspective. Following are provisional frames that draw on the first two sections of this essay's analysis and a third new one to help illustrate what the direction of an appropriate model of scientific literacy in the postgenomic era using ANT might look like.

Frame I: Overcoming Nature and Embracing Pedagogical Nonhumans

The Pacific salmon exemplifies a perfect case study of one of the most important barriers for developing student/civic skills that would allow for a more democratic relationship to form between nonhumans such as the AquAdvantage® Salmon and the public. As I have argued earlier, the plight of salmon has been established as a scientific problem that, in turn, blocks it from being a common concern in a number of important ways. Such a binary between nature and culture and specifically who gets to work on problems in nature *and* how is foundational to any democratic rethinking of teaching and learning of science in the postgenomic era. As the epistemological tool of environmental history has helped us discover and took us one step toward thinking more like an ANT, nonhumans, if nothing else, aid in co-constructing our social realities in a variety of ways. The salmon, dam, army corps of engineers, and hydroelectric power are all actors inextricably linked to an actor network that only makes sense if one begins to see the power relations *and* constructive qualities of each relationship at both the individual and collective levels. To better represent such a reality, science education combined with social studies perspectives (as ANT holds, the social and the scientific should not be treated as two separate areas of study), I am suggesting, should begin to treat nonhumans such as salmon as a site of ANT enquiry. One practical way to model such a goal is by building actor network maps in science education settings that draw upon environmental history to excavate obscured and alternative relations that are a product of educated amnesia to past human and nonhuman relations. That is, instead of salmon being understood solely through modern scientific lenses such as their species classification, biological mechanisms, or molecular composition, the nonhuman needs to also be engaged as a pedagogical actor capable of contributing to the learning process of a more capable network epistemology, especially in cases where science and technology have intervened deeply into forms of life.

For instance, one question educators could ask as a point of departure for creating actor network maps is: what are some of the most influential actors that make up the composition and character of relations between Pacific salmon and the AquAdvantage® Salmon? By framing the research question with a focus on the number and composition of relations, a complex and highly educative map can be constructed through a process of collective (or individual) enquiry and experimentation in both formal and informal educational settings (science and social studies classes, community research groups, etc.). A list of potential actors to research this question (though by no means exhaustive) with the aid of an environmental history lens could comprise something like the following: research on fish science policies (past and present) promoted and enforced by state and federal governments, the FDA guidelines for receiving and reviewing a "new animal product" license and what actors make such a designation possible, dietary and cultural importance of salmon to American Indian groups in the Oregon country compared to the contemporary importance of GE salmon in US food culture, the relationship between sea lice from pen-raised Atlantic salmon and wild Pacific salmon runs, or research on the relationship between deforestation, urban development, and the need for alternative ways of producing fish from a declining wild population.[15] Any one of these lines of investigation is a fertile ground for producing an actor network map that can both utilize science as a tool for enquiry and also recognize and draw upon the pedagogical potential of the nonhuman. From such a process, following the basic principles of scientific investigation such as developing a research question, experimentation, collecting and analyzing facts, observing co-constructive relations, and the public sharing of research results with community members, students have the opportunity to start enquiry from scratch (and from below) on relevant community problems.[16] However, in line with one of ANT's fundamental tenets, creating actor network maps should not utilize the modern categories of "nature" and "culture" as epistemological restrictions in the construction of new knowledge using ANT (Latour 2005).[17]

Once such categories are imposed we are right back where we started from, and suddenly the problem of salmon becomes an issue for experts who work out in nature and return to tell communities what to do and how to think about public problems. The established and accepted relations with GE fish are chaperoned in a way that prevents them from entering democratic politics. The practice of ANT is far more open. ANT requires that facts be discovered and constructed through processes of enquiry and experimentation that an actor network under investigation demands (Latour 2005). In this way, it is possible to tap into the pedagogical potential of nonhumans and their relations with humans that make up many of the most salient social facts constituting many important aspects of our daily lives, with GE food being one of a multitude of such controversies. See figure 4.1 for an example of actors that are connected to an actor network of salmon. Here, the salmon itself represents the nodal point for the various nonhuman and human actors involved in the life of salmon—each actant—for example, the FDA—designates a constellation of humans and nonhumans involved in the controversy surrounding GE fish, and GE food in general.[18] From a modern scientific approach, however, the salmon would represent the bifurcation of society (the top half of the salmon) and culture (the bottom half), leading to enquiry that fails to connect the entirety of actors involved since the human and nonhuman worlds can never intermingle and

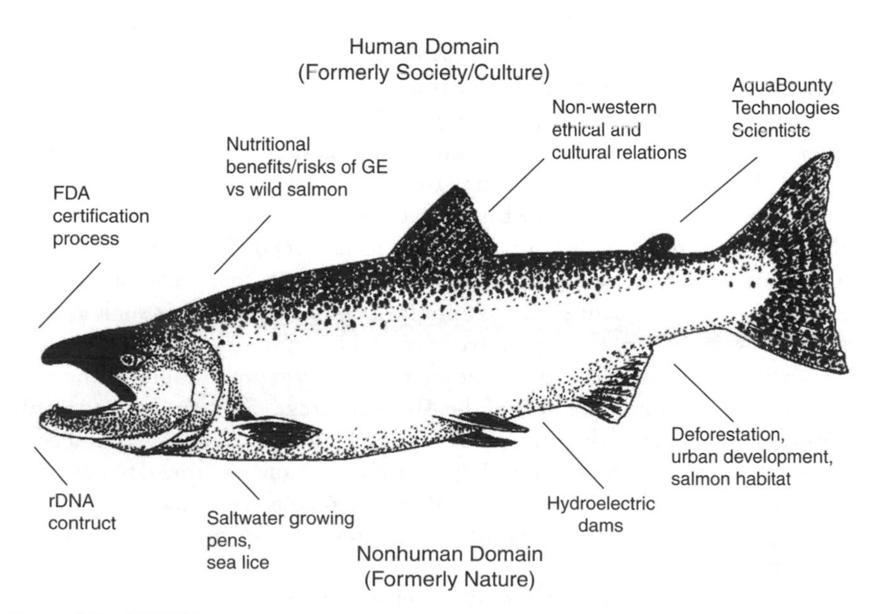

Figure 4.1 ANT Salmon

enter into productive democratic relations in the modern model of science. The figure of the salmon and the actors that connect it to social and scientific life provide a starting point for an ANT epistemology that truly embraces the pedagogical nonhuman—in this case, the salmon.

Frame II: Civic Technoscientific Literacies

As I pointed out in the second section, the dominant model of scientific literacy being promoted at both the policy and the curricular level in K-12 science education reform is tied to a growing economic need that requires a higher domestic output of STEM skills and literacies from the education system in the United States. AquaBounty Technologies' scientific construction of their GE fish in the FDA report represents the type of scientific literacy, I argue, that is most sought after by a number of powerful stakeholders and policymakers in the current neo-*Sputnik* reform movement. In combining biological and molecular with engineering literacies, the new science education standards set up the primary use-value model for teaching and learning science directly in line with technoscientific literacies. What's more, the public construction of AquaBounty's application for a "new animal product" under review by the FDA also reinforces the view that what the scientists in the laboratory are up to has nothing to do with social concerns; in fact, their own scientific evidence supports the view that the GE fish of their making create no ecological or human risks—thus reassuring the public that the nature/culture divide is quite secure (FDACVM 2010).

Opposed to this neo-*Sputnik* model of scientific literacy, ANT provides an alternative that does not take its cue from the global economic concerns of the US government and those of one of the largest biofarming corporations in the world. What I am suggesting should be turned to in ANT for developing a more democratic model of scientific literacy is a recognition of how technoscientific literacies are already enlisting humans and nonhumans into co-productive relationships. That is, just as the National Academies' new science education standards has recognized, the real practice of science is no longer attached to the myth of a solitary man in a white lab coat working diligently on a complex problem by light of a candle. In fact, what the authors of the new science education standards have responded to is the fact that science is a co-constructive process that involves scientists across disciplines and nonhumans such as gene sequencing software in order to solve intractable mysteries of nature such as the human DNA code. This aspect of technoscientific production can be seen clearly in the nonhuman example of the AquAdvantage® Salmon, an entity only possible through the collaborative work of genetic information from two species of fish, a biofarming corporation, its bioengineers, and an amenable governmental policy that allows such a chain of actors to come together and flourish, and a science education classroom that exonerates students from learning about such entangled networks.

For instance, one of the most important constituent parts of the AquAdvantage® Salmon (the rDNA construct created from an Ocean pout and Chinook salmon) involves two nonhuman actors that contribute to the construction of the GE fish created by AquaBounty Technologies. The genetic information, along with the technologies and procedures employed by bioengineers, create an actor network that would not have been possible without the input and qualities of two different nonhumans' genetic information. Here we have what science and technology studies theorists have called "science in action." This is the base of what I am suggesting needs to be a civic technoscientific literacy: a way of learning and teaching about science and technology that recognizes the co-constructive role that nonhumans play in the production of scientific knowledge and practice. What is different from the technoscience literacy of the genetic lab, however, and again where ANT plays a significant part, is that the teaching and learning of technoscientific literacies also need to open up moral and political questions that are expunged through current approaches in laboratories and bioscientific research facilities. In other words, if education is going to take up the role of cultivating technoscientific literacies (not an inherently bad thing since it reflects the actual practice of science today more accurately), it also needs to put on the pedagogical table the question of whether or not communities and the public writ large want to splice genes as a food production model of the future. As it stands now, and as my analysis of the new science education standards alongside the AquaBounty report reviewed by the FDA, this decision is not open for debate—or, at least, in any real sense of the word. A civic technoscientific literacy thus would offer both parts of the whole instead of isolating one for the benefit of bioindustrial growth. The moniker of "civic," in other words, designates an opening into the reality of technoscientific literacy that involves moral and political questions that have yet to be a part of the discussion of revamping educational conceptions of scientific literacy in our current era. Educational practices that utilize actor network theory as an epistemological tool in the creation of actor network maps, for instance, should also be working with a civic technoscientific literacy as a basis for tracing chains of actors and articulating in clear terms their composition and character.

As figure 4.2 illustrates, civic technoscience literacy involves tracing the relations *and* composition between actors. For instance, the link between wild salmon and sea lice is only explainable through the network composition of GE salmon/growing pens/rDNA constructs/aquafarming corporation.[19] What a civic technoscience literacy would emphasize in this actor network is a more accurate account of a controversy, in this case the extinction of a wild salmon run. GE salmon, understood through a civic technoscience literacy, do not exist as isolatable entities that can be purified through technologies, such as growing pens. In focusing on the co-constructive quality of science in action (especially biotechnological expressions), a more factually precise picture comes into the frame. Namely, that the controversy of salmon extinction and the politics that surround it must include an account of the relational and co-constructive chain of actors, as each one plays a constituent role in the manufacturing of sea lice explosions within the migratory waterways of wild salmon.[20] The road to democratic accountability and involvement becomes less opaque through

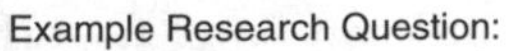

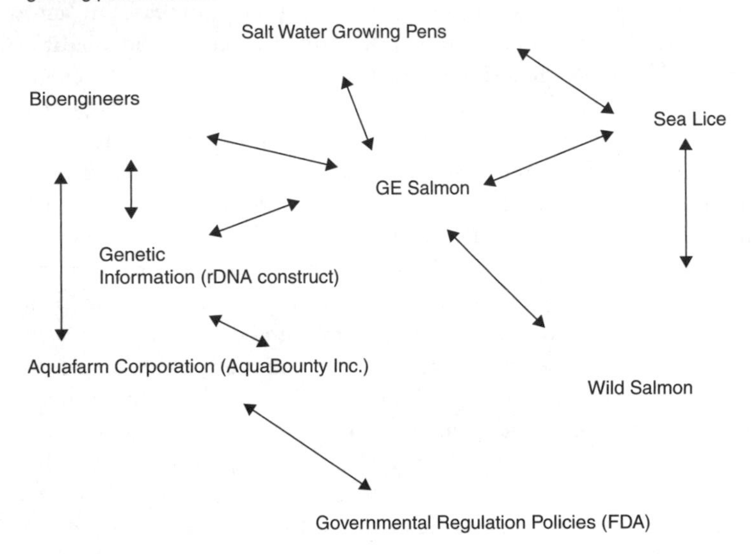

Figure 4.2 Civic technoscience literacy: Tracing co-constructive relations.

such a practice of scientific literacy because the GE salmon debate is no longer understood through the practice of purification and blindness to human–nonhuman networks. As such, the debate for a "new animal product" license, reframed through civic technoscience litearcies, must now involve a whole host of actors that can speak as their compositional and relational character is no longer held silent.

Frame III: Democratic Education in the Postgenomic Era

In the previous two frames of transition and throughout the entire journey of our salmon, what has become apparent is that democratic education, given its modern inheritances, is utterly ill-equipped for dealing with serious public problems of the present and those of the not-so-distant future. If nothing else, what I have been arguing is needed are educational approaches that are attuned to deal in productive and democratic terms with problems such as GE salmon and its attendant controversies. What is clear is that science education (and education in general) is failing to prepare citizens, communities, and teachers with the tools needed to make sense of and include everything and everyone that comprise our common social realities.

What Latour and other theorists and activists who are rethinking democratic theory from the standpoint of the common (Hardt and Negri 2000, 2005, 2008; Shiva 1998, 2006) have pointed out is the need to cultivate new habits and skills in individuals and collective groups that can emerge from a common ethic. In applying ANT as an educational practice, epistemological landscapes arise that can potentially lead to a stronger understanding of the ways the commons are being involved in processes of capitalization and cultural and economic enclosure. Such a location for producing knowledge is important because, as my analysis of GE salmon using ANT attempts to show, practices and governing laws foreclose on the development of habits and literacies that can frame something like GE food debates within larger networks of power. In utilizing the concepts of nature and culture to purify reality, bioindustrial interests prevail over the common forms of life it requires to grow as a productive model. Part of developing educational habits reoriented toward a common ethic, therefore, requires that educators, learners, and community members work with epistemological tools that can bring nonhumans into democratic life by creating a more coherent vision of the ways human and nonhuman life are involved in economic and cultural systems based on domination and extractive ethics. As Vandana Shiva has argued in her work, "patents on life and the rhetoric of the 'ownership society' in which everything—water, biodiversity, cells, genes, animals, plants—is property express a worldview in which life forms have no intrinsic worth, no integrity, and no subjecthood." Such a view and ethic towards life also "holds the anti-life philosophy of those who, while mouthing pro-life slogans, seek to own, control, and monopolize all of earth's gifts and all of human creativity. The enclosure of the commons that started in England created millions of disposable people. While these first enclosures stole only land, today all aspects of life are being enclosed—knowledge, culture, water, biodiversity, and public services such as health and education" (Shiva 2006, 3).

Currently educational practices that are driven by the Race to the Top and race against other advanced industrialized nations that have surpassed the economic capability of the United States in biotechnological and bioscientific fields only make it more difficult to develop a scientific literacy that can deal adequately with pressing ecological and social crises. Furthermore, instead of creating a pedagogical bridge to common life, a place where habits and skills can be cultivated that recognize network life and the power relations that constitute it, scientific literacies built on the understanding that "nature" is an extractive and controllable object only advance crises that have reached their tipping point without a tenable exit strategy. Global climate change, widespread species extinction, deforestation, famine, access to basic medicine, and shrinking fresh water supplies across the planet all point to the need for ways of thinking and acting that hold the common more important than high STEM scores or greater investments into human capital through schooling.

The final lesson to be drawn from ANT as an epistemological tool for creating a more appropriate form of scientific literacy is a focus on commonality instead of individuation, separation, and proprietary fears. In creating actor network maps as a pedagogical exercise, another potential benefit is a civic ethic of commonality that modern scientific approaches fail to produce and, in fact, resist in favor of enclosure by turning nonhumans into private, non-democratic goods. That is, through the practice of researching and tracing actor network relations, students and teachers are opened up to a highly dependent (and actively constructed) universe where the well-being and health of salmon are also related to their human counterparts. What can unfold in the act of thinking like an ANT is a view of a set of power and co-constructive relations that reflects back from the nonhuman world into the human, showing that in fact the AquAdvantage® Salmon is just as much a part of us as we are of it. It is this common ontological lesson that ANT has the potential to render as a pedagogical tool for learning and teaching about life in the postgenomic era. From such a standpoint, a viable politics of the future could be created, one where educational habits can be formed that recognize the stakes of something like the GE food debate to be life itself.

If our salmon has taught us anything, it is that the common waters that nonhumans and humans share and have shared for millennia need to be the horizon from which future educational approaches are created. One way to begin to see the ocean that surrounds us is to follow the ANT as it chases the salmon. Through paths such as these, network life and epistemologies for detecting the multitude of lines it can follow open up a whole new world of understanding and hopefully learning and being in the world. Thus, the stakes of developing an alternative scientific literacy to those that call forth the AquAdvantage® Salmon are much higher than the health of a quickly growing economic sector. It is about making sure that the common waters hold on to the depleting diversity of nonhumans that teach us more than ways to increase potential quarter earnings or how to create cost-effective and risk-laden food products. With the yearly closure of untold salmon runs, many of the most fragile actor networks are quickly shutting down. Education in the postgenomic era should be about keeping these networks open as well as thinking of ways to make them healthy and to come up with methods for including their participation in democratic life.

PART III

Biological Citizenship in a Flat World: Governmentalities of Optimization and Their Alternatives

CHAPTER 5

The Biomedicalization of Kids: Psychotropic Drugs and Biochemical Governing in High-Stakes Schooling

> Like all new technologies, cognitive enhancement can be used well or poorly. We should welcome new methods of improving our brain function. In a world in which human work spans and lifespans are increasing, cognitive enhancement tools—including the pharmacological—will be increasingly useful for improved quality of life and extended work productivity, as well as to stave off normal and pathological age related cognitive declines. Safe and effective cognitive enhancers will benefit both the individual and society.
>
> Statement by neuroscientists in *Nature* article "Towards Responsible Use of Cognitive-Enhancing Drugs by the Healthy."
>
> —December 2008

For many, corporeal punishment is thought of as a practice relegated to a bygone age, a barbaric ritual of the past that has disappeared almost completely from daily life in US schools. Returning to the etymology of the term, however, suggests something different. After all, corporeal punishment is a type of corrective action that has the intent of getting the physical body of individuals in line with normalized understandings of acceptable and desirable behavior within institutional settings. Paddling, spanking, and thwacking students' hands with a ruler all had one clear purpose in the school: to inflict pain on the human body as a means for adjusting the behavior and disposition of pupils within a permitted range of behavior. In other words, corporeal punishment can also be understood as a type of bodily pedagogy committed to control and regulate students within the disciplinary structures of school that attend to the educational needs of the state, the economy, and cultural norms.

Of course, Michel Foucault famously charted the historical development of such types of disciplinary techniques used throughout the history of modern institutions to control human bodies in accordance with forms of sovereign power associated with the church, school, military, factory, and asylum.

Understanding corporeal punishment from a genealogical perspective as Foucault did, it is quite clear that practices of disciplining the body in schools have not disappeared. Rather, they have been folded into more subtle and complex systems of bodily control and pedagogy informed by neoliberal rationalities and practices concerned with the efficiency, value maximization, optimization of performance, and measurability of students' bodies. However, in the age of biocapitalism, technoscientific advances in biopharmaceutical drugs have ushered in a new regime of bodily control that co-articulates with existing neoliberal forms of educational governmentality concerned with optimizing the human capital stock of the nation. Forms of educational discipline and control in the age of biocapitalism, I argue in this chapter, therefore need to be "situated in a certain 'political economy' of the body: even if they do not make use of violent or bloody punishment, even when they use 'lenient' methods involving confinement or correction, it is always the body that is at issue—the body and its forces, their utility and their docility, their distribution and their submission" (Foucault 1995, 25).[1]

This chapter examines the terrain of what Foucault calls the "political economy of the body" by examining the proliferating use of psychotropic drugs as a behavior management technique in schools. In order to fully understand the deeply troubling growth trend of psychotropic drug use in schools over the past 20 years, I argue that it is necessary to interpret such a growth pattern within the genealogical arc of types of bodily control in schools that began with corporeal punishment practices to current ones that utilize biochemical technologies for regulating bodies within educational populations. Put differently, drugs such as Ritalin, Concerta, and Adderall do not represent a rupture in the ways bodies in schools have been traditionally regulated and disciplined. Instead, I argue in this chapter that psychotropic drugs represent a deeper level of control that is achievable through the advancement of biopharmaceutical research and development, *and* how the public (most importantly parents, teachers, and school administrators) think about both "mental disorders" and treatments for the "sick" or "diseased" body in educational settings.

One of the larger questions this chapter seeks to understand is how the spank of the paddle employed in the industrial age of schooling has been replaced with tools capable of neurologically reordering children's bodies in the biocapitalist age. Yet I also want to ask how new biotechnologies such as psychotropic drugs participate in a unique set of neoliberal strategies and practices that target the body of students so as to achieve maximum efficiency and value extraction from student populations. In other words, the alarming rate of growth in psychotropic drug use in schools, I am suggesting, is not only a question of technological advancement (i.e., the ability to alter one's neurological pathways through biochemical intervention), but also a particular neoliberal form of governmentality that has normalized the way individuals and institutions think of the body and, in particular, how to make one's self more valuable and productive within a highly competitive and market-driven educational culture.[2] Racing to the top, now includes new technological advancements like "cognitive-enhancers" to achieve global dominance in education and economic superiority.

What also cannot be lost from sight in examining the relationship between the political economy of the body and its regulation through psychotropic drugs in schools is how one of the most powerful economic actors in the twenty-first century bounds both the field of knowledge and practice of mental health: the biopharmaceutical industry. As Nikolas Rose has convincingly shown in his work, psychopharmacueticals have become one of the most powerful sectors in the global economy. Different from the first part of the twentieth century when mental health problems were largely understood in eugenic terms as something detrimental to society, a sign of population degeneracy, psychological health is now a booming area of market growth. Rose (2007) notes:

> In the eugenic age, mental disorders were pathologies, a drain on a national economy. Today, they are vital opportunities for the creation of private profit and national economic growth. Indeed the profit to be made from promising effective treatment has become a prime motive in generating what counts for our knowledge of mental disorders. Over the last twenty years or so, in advanced industrial societies of Europe and North America, psychopharmacology has carved out a very significant market. Over the decade from 1990 to 2000, the psychiatric market increased in value over 200 percent in South America, 137 percent in Pakistan, 50 percent in Japan, 126 percent in Europe, and a phenomenal 638 percent in the United States. (209)

However, within this pattern of explosive economic growth, Rose (2007) also points out the following:

> In both the United Kingdom and the United States, one key growth area has been the new SSRI type antidepressants, increasing by around 200 percent over the decade, with a related, but smaller, decline in prescriptions for anxiolytics. A further feature, and one that has attracted much controversy, has been the rise in prescription of psychostimulants, notably Ritalin (methylphenidate) and Adderall (dexamphetamine), for treatment of Attention Deficit Disorder in children. From the mid 1980's to the end of the twentieth century there was a remarkable growth in the diagnosis of this condition and in the use of these drugs. This was most marked in the United States, where prescribing rates rose eight-fold in the decade from 1990 to 2000. (210)

As I discuss in this chapter, the growth rates of psychostimulant diagnoses and use have continued to grow at an exponential rate in the following decade (2000–2010) as well.

The story of the attention deficit/hyperactivity disorder (ADHD) epidemic in the United States, therefore, cannot be told outside of the powerful regime of the biopharmaceutical drug industry. Market growth and diagnoses rates are inextricably linked. It is not an accident that just as the psychopharmacology industry began to grow, so too did the need to find more patients (customers) in need of treatments to sicknesses they perhaps were unaware existed and that their bodies were afflicted with. With this said, however, I want to make one point clear about my analysis in this chapter: I am not suggesting that

ADHD is a false or contrived health condition; I am simply not qualified to make such a judgment. What I am arguing however, and what seems undeniable given the data showing a skyrocketing increase in both ADHD diagnoses rates and prescription drug use by children over the past 25 years, is that bodies in schools are increasingly being subjected to a co-articulating economic and medical form of biopower that has no historical precedent. On the one hand, a whole knowledge regime has been built around being able to diagnose and recognize the pathology that has come to be known as ADHD, which is also connected to one of the most powerful and fastest growing industries in the global economy. On the other hand, schools in our high-stakes neoliberal moment, as I have argued in previous chapters, can perhaps best be understood as spaces of biocapital production and thus places where educational life is ever more becoming the target of an expanding range of sophisticated technologies of control. These two tensions, the growing influence of the biopharmaceutical drug industry on how humans understand health in relation to their selves *and* emerging practices for optimizing educational life, are the target of my analysis and not the validity of ADHD as a mental health condition.

In order to see how these tensions are playing out under the neoliberal restructuring of education currently taking place in the United States, I begin with an examination of the most recent available data on ADHD diagnosis rates and psychotropic drug use for children in the United States. From here I interpret the staggering growth rate of ADHD diagnosis rates by focusing on the ways in which the pharmaceutical industry has become an important pedagogical actor in constructing ADHD knowledge and practices that have been adopted in school settings across the United States. To draw out how the pharmaceutical industry is engaged in a type of public pedagogy, I use Nikolas Rose's concept of biological citizenship to examine how the collaborative educational enterprise between the pharmaceutical industry and institutional actors such as the National Institute of Mental Health (NIMH) and the Department of Education (DOE) helps shape knowledge and practices surrounding ADHD children in schools. It is particularly important to focus on this pedagogical relationship since it is highly influential to how teachers, parents, school psychologists, counselors, and administrators confront something like the problem of a potential ADHD student. As I argue in this section, one important aspect of biological citizenship that has emerged in schools around ADHD is the establishment of a biological or genetic basis for the disorder that has deep implications regarding how students' bodies are introduced and enmeshed within new practices of control.

Next, I look at the ways in which a genetic framing of biological citizenship related to ADHD in schools elicits and normalizes a type of biochemical governance over children's bodies. That is, once a genetic explanation forms the basis for understanding ADHD in school settings (largely through the pedagogical efforts of big Pharma and other institutional actors), practices centered on psychotropic drug intervention are also involved in a process of naturalizing biochemical control as the most rational treatment response. In mapping the interacting rationalities and practices that constitute what I am calling biochemical governance, I look to two important sources: the DOE's teacher's manual for identifying ADHD in students and George DuPaul and Gary Stoner's leading guide book on ADHD intervention strategies for teachers and school personnel *ADHD in the Schools*. These two sources are representative of a model of governmentality that has emerged through the uniting of the dominant knowledge regime on ADHD as a biological or genetic problem and the attendant intervention practices that rely on psychotropic drugs for regulating children's behavior and disposition in schools.

Finally, I consider how the emergence of biochemical governance associated with the ADHD epidemic in the United States co-articulates with current neoliberal educational imperatives focused on improving and optimizing the quality of human capital stock in this nation. Such a co-productive framework, I suggest, signals an important new dimension in the biopolitics of education today. Biotechnologies such as Adderall and Ritalin provide the means to achieve an incredibly deep level of control over educational subjects. It is not only the habits and desires of students that are targeted, but the very neurochemical composition of children that is at stake. Here, the convergence of biocapitalism and education is evident in such a model of biological control since student optimization can only be achieved through biocapitalist technologies (psychotropic drugs) and neoliberal educational environments that call for greater and greater degrees of regulation and discipline over the body of the student. Given this confluence between neoliberal educational imperatives and biocapitalist technologies of control, we must begin to ask where strategies of resistance might emerge. How, in other words, can alternative models of biological citizenship be produced outside of knowledge and practice regimes such as the governmentality model that regulates ADHD bodies in schools? In exploring this question, I look at the food sovereignty and autonomous school movement in Southern Mexico for an example of an alternative model of biological citizenship, specifically a model of biological citizenship that is defined and produced in resistance to neoliberal development policies that are involved in biocapitalist expansion and territorialization of both corn and people. First, however, it will be helpful to get acquainted with some of the most recent figures on ADHD diagnosis growth in the United States.

Teaching Sickness: Big Pharma as Public Pedagogue

The most current data available on growth rates in ADHD diagnoses and treatment in the United States are alarming. The following are the most salient figures produced by two recent Center for Disease Control and Prevention (CDC) studies (2008, 2010) on ADHD in the United States:

1. Approximately 9.5% or 5.4 million children 4–17 years of age have ever been diagnosed with ADHD as of 2007.
2. The percentage of children with a parent-reported ADHD diagnosis increased by 22% between 2003 and 2007.
3. Rates of ADHD diagnosis increased an average of 3% per year from 1997 to 2006 and an average of 5.5% per year from 2003 to 2007.

4. Boys (13.2%) were more likely than girls (5.6%) to have ever been diagnosed with ADHD.
5. Rates of ADHD diagnosis increased at a greater rate among older teens compared to younger children.
6. The highest rates of parent-reported ADHD diagnosis were noted among children covered by Medicaid and multiracial children.
7. Prevalence of parent-reported ADHD diagnosis varied substantially by state, from a low of 5.6% in Nevada to a high of 15.6% in North Carolina.

Other important findings from the recent CDC studies on ADHD diagnosis rates show that "among older teens, parent-reported ADHD diagnosis increased by 42% and among Hispanic children it increased 53%, suggesting that the epidemiology of ADHD in the United States may be changing. Hispanics have historically had some of the lowest rates of ADHD. Although still lower than non-Hispanics, the relatively large magnitude of change in this study suggests that there may have been a shift in the cultural perception of the disorder or a change in the diagnostic practice for Hispanics" (CDC 2011). In the most recent CDC study, the percentage of children ever diagnosed with ADHD increased from 7% to 9% from 1998–2000 to 2007–2009. Also, differences in ADHD diagnosis between ethnic and racial groups narrowed during the period of 1998–2009. For children with family income less than 100% of the poverty level, ADHD prevalence increased to 10% and 11% for children with family income between 100% and 199% of the poverty level (Akinbami et al. 2011).[3]

By any objective measure, it is clear that ADHD diagnosis rates are escalating at a rapid pace in the United States. Mirroring the growth rates of ADHD diagnosis in the United States, not surprisingly, has also been a gigantic boom in the production of amphetamine prescription drugs such as Adderall, Ritalin, and Concerta. As Lawrence Diller (2011) has pointed out recently using data from the United Nation's International Narcotics Control Board (2010), the United States "which makes up four percent of the world's population, produced eighty eight percent of the legally prescribed amphetamines." Just as alarming, Diller (2011) also notes that according to a study done by the US Drug Enforcement Agency, drug use of prescription amphetamines has multiplied tenfold from 1996 to 2009. In terms of worldwide production of amphetamine prescription drugs, the United States manufactured 98.9% of the global supply (6675 kg in 2009), while France, the only other producing country, came in with less than 2% in 2009 (United Nation's International Narcotics Control Board 2010).

How should parents, teachers, staff, and school administrators as well as the general public make sense of such a growth trajectory in the diagnosis rate for ADHD and its treatment through amphetamine-based drugs such as Adderall, Ritalin, and Concerta?[4] Indeed, as others have suggested, it is quite accurate to refer to ADHD diagnosis as an epidemic in the United States (Diller 1998, 2011; Breggin and Breggin 1998; Breggin 2001). Furthermore, what should these same groups make of the fact that psychotropic drug use in schools has been increasing in tandem with neoliberal accountability measurement devices such as merit pay incentives linked to value-added metrics that put even more pressure on teachers and students to perform at high levels within a classroom succumbing to an increasingly narrow definition of educational achievement? Placing the above data on the exploding rate of ADHD diagnosis and treatment in the United States alongside educational policies such as NCLB and now Race to the Top—both of which are built upon highly restrictive and standardized definitions of educational performance and achievement as well as market-driven punitive schemes—an unmistakable coevolutionary pattern appears. Given the dual trends of ADHD diagnosis and acceleration of punitive accountability practices in classrooms, the social ecology of schooling produced through neoliberal policies and practices creates intense environmental stressors for students, teachers, parents, school personnel, and administrators. As such, the need to regulate bodies in schools in line with other technologies of control such as value-added metrics (the subject of chapter 3) or competitive enrollment policies, for example, converges with the needs of the pharmaceutical industry to continually seek (and create) an accessible consumer base that must fit into a stricter set of parameters that demark acceptable and standardized behavior in the classroom. In short, schools under strengthening neoliberal forms of governmentality imbued with rationalities of efficiency and value maximization are incubators for producing "abnormal" behavior in individuals, especially when the criteria of abnormal behavior is largely influenced by pharmaceutical corporations that hold the biochemical solution to better educational performance (Breggin 2001).

In order to make better sense of the ADHD epidemic in the United States, especially as it relates to neoliberal structures that continue to shape the social ecology of schools, I suggest that the problem of overdiagnosis and treatment needs to be thought about in terms of what Nikolas Rose has called biological citizenship. Biological citizenship is a concept that Rose has developed looking at how individuals' and groups' understanding of self is transforming under biomedical and biotechnological regimes of knowledge and practices in society. Rose's notion of biological citizenship is useful because it helps elucidate how biochemical approaches for controlling students' bodies have become a normalized practice in schools, and is indicative of a reflexive relationship that has emerged between a powerful new knowledge regime (pharmaceutical industry/American Psychiatric Association/DOE) and individuals and groups that are increasingly coming to understand health and care of self through such regimes. Rose (2007) describes how such a pedagogy of the self, which is the basis of biological citizenship formation, operates:

> The new psychiatric and pharmaceutical technologies for the government of the soul oblige the individual to engage in constant risk management, to monitor and evaluate mood, emotion, and cognition according to a finer and more continuous process of self-scrutiny. The person, educated by disease awareness campaigns, understanding him-or herself at least in part in neurochemical terms, in conscientious alliance with health care professionals, and by means of niche-marketed

> pharmaceuticals, is to take control of these modulations in the name of maximizing his or her potential, recovering his or her self, shaping the self in fashioning a life. (223)

It is in this nexus between the knowledge regime of the pharmaceutical industry working hand in glove with government regulatory agencies, and how individuals and groups learn to think about their health and illnesses that the ADHD diagnosis epidemic plaguing schools and US society in general must be understood. One way to get a better sense of how the relationship between the pharmaceutical industry and the educational subjects has developed is to look at the amount of money drug companies are dumping into advertising and marketing research to create and maintain an effective public pedagogy about their products. In what has been aptly named "disease mongering" in a recent *British Journal of Medicine* article that looks at how pharmaceutical education campaigns flood the public with information on health conditions such as erectile dysfunction, irritable bowel syndrome, or male pattern baldness, teaching people how to think about illness is big business (Moynihan Heath, and Henry 2002).[5]

It is no accident that the watershed year (1980) for the pharmaceutical industry coincided with the birth of neoliberalism as state doctrine in the United Kingdom and the United States. As Marcia Angell (2004), former editor in chief of *The New England Journal of Medicine*, notes in her devastating analysis of the pharmaceutical industry, "Before then, it was good business, but afterward, it was a stupendous one. From 1960 to 1980, prescription drug sales were fairly static as a percent of the U.S. gross domestic product, but from 1980 to 2000, they tripled. They now stand at more than $200 billion a year" (3). In 2009, this figure grew to over $300 billion (Berkrot 2010).[6] One of the chief reasons for such a remarkable growth bonanza in the pharmaceutical industry was the passing of the Bayh–Dole Act that "was designed to speed the translation of tax supported basic research into useful new products—a process sometimes referred to as 'technology transfer'" (Angell 2004, 7). The net result of legislation such as the Bayh–Dole Act was a massive transfer of public money directly into the biotech and the pharmaceutical industries via governmental agencies such as the National Institute of Health (NIH). With legislation such as the Bayh–Dole Act, public university research and development were permanently married to corporate interest and private profit. Along with this meteoric ascent to be the most profitable industry in the nation, drug companies recognized and mastered a complex system of public relations and advertising to educate the population about their wares. If the Bayh–Dole Act and other sets of legislation such as the Hatch–Waxman Act that extended monopoly rights for brand-named drugs for companies paved the regulatory way during the Reagan era, the other part of the equation the pharmaceutical industry needed to balance was ways to teach the populace how to recognize illnesses in themselves and how to treat it.

One of the most well-known strategies pharmaceutical companies have deployed for "educating" the public is its involvement with the group Children and Adults with Attention Deficit /Hyperactivity Disorder (CHADD). As Peter Breggin (2001) has painstakingly documented, "Ciba's [maker of Ritalin and now called Novartis] best partner—is CHADD, an organization of parents whose leadership supports biological explanations and drugs for the problems that they have with their children" (233). Breggin (2001) goes on to point out that "CHADD was formed in 1987 as a parent-based organization and quickly became a major player in mental health politics. Its leaders are parents who have strongly embraced ADHD diagnoses for their children and treat them with Ritalin or other drugs . . . A substantial portion of CHADD's support comes from CibaGeneva Pharmaceuticals" (234).[7] What CHADD and other groups like it represent is the coercive landscape in which parents, teachers, administrators, and the general public find themselves when it comes to making decisions regarding student mental health and normal classroom behavior.

The pharmaceutical industry's health education enterprise, however, has been developing for quite some time. Ilina Singh's research on the evolution of drug companies' media campaigns in the mid part of the twentieth century, for instance, demonstrates how the pharmaceutical industry has long recognized that the most effective health education for the public is one that targets existing social and cultural norms in their marketing of a new drug. In her analysis of post-World War II educational magazines and books on childrearing aimed at mothers, Singh (2002) points out how

> mothers trained in the wisdom of the permissive era were likely in need of some education in face of these new ideals of childrearing. Good mothering was now dependent on discipline, and discipline had an important relationship to "normalcy." Articles on discipline-related areas, such as obedience and aggression were heavy with talk about normative levels of (boys') aggressiveness, impulsiveness, and conformity. Such articles emphasized that mothers should be able to make a distinction between normal and abnormal behaviors, but confirmation of this distinction and intervention of most any kind was the job of the expert. (585)[8]

What Singh's analysis shows is that drug companies, in this case the makers of Ritalin which first appeared in the US market in the 1950s, targeted "educational" audiences they most desperately needed to accept notions of normal and abnormal behaviour, in particular, how their drug can adjust boys' "naturally" aggressive behavior and help mothers raise their children to be productive adults, a task designated by the mother's role as primary caregiver. Indeed advertising spending (reported as "educational activities" by drug companies) has grown rapidly over the years. In a recent study looking at the amount spent on advertising and other public relation expenditures, the pharmaceutical industry spent double the amount on advertising ($57.5 billion in 2004) as it did on research and development (Gagon and Lexchin 2008). Clearly such lopsided spending indicates how important it is for the pharmaceutical industry to develop and maintain a public education regime that teaches the population how to think about disease and treatment, so much so that making new drugs is a subordinate imperative to making new customers.

In addition to the targeting of mothers and families through front groups and media campaigns, there are also more direct methods that pharmaceutical companies pursue in "educating" key markets about their products. Given the fact that most ADHD diagnoses begin with teachers, school administrators, nurses, and school psychologists, drug companies have developed a multitiered approach for coaching professional educators on recognizing ADHD signs in students and what to do about it when the telltale behavioral patterns emerge in students. Pharmaceutical companies such as Novartis (now makers of Ritalin) and Teva and Barr Pharmaceuticals (makers of Adderall), for example, concentrated their educational campaign efforts on stand-alone websites designed to be resources for teachers, provided toll-free help lines for teachers and nurses, and collaborated with groups such as the National Association of School Nurses to run "a nationwide campaign in which 11,000 school nurses were provided with a resource kit containing information on ADHD, its treatment, and various support organizations" (Phillips 2006, 0434). Such direct types of corporate coercion (one might also call it a type of perverse pedagogy) utilized by pharmaceutical companies are not shots in the dark attempting to pick up a few more customers. Rather, companies such as Novartis and Teva Pharmaceuticals know precisely that teachers are one key group in which their message must resonate and, most importantly, become the operational knowledge by which behavioral and discipline decisions are made in constructing biological citizenship in the school.

In a recent statewide study of general and special education teachers, it seems that pharmaceutical education campaigns targeting teachers and school personnel are having the desired effect. In a random sample study of 400 teachers in the state of Wisconsin (200 general and 200 special education), a series of surveys revealed that "teachers' opinions about the effect of stimulant medication on school-related behaviors were generally positive" (Snider, Busch, and Arrowood 2003, 46). Though perhaps most striking, in that it validates the drug companies' strategies for targeting teachers through educational campaigns, is one of the study's major findings: "Two thirds of the teachers in this study indicated that teachers were the first to recommend that a child be evaluated for ADHD. This confirms previous research indicating that as many as 40% to 60% of initial referrals come from teachers." The authors then go on to point out that "this is also consistent with the perceptions of other school professionals. When speech and language clinicians and school nurses were asked to identify the professional who first recommended assessment for ADHD, 74% of speech and language clinicians and 80% of school nurses said that teachers were the first ones to suggest an assessment for ADHD" (Snider, Busch, and Arrowood 2003, 53). It would not be outlandish to claim that it is highly probable that pharmaceutical companies have been way out in front of such studies in their market research on how to cultivate and shape such a health culture in schools.

Another important aspect to the actor network involved in making teachers de facto ADHD clinicians has been the scientific framing of the disorder by governmental agencies such as the NIMH as a biological-neurological disease (Dillard 1998, Breggin 2001). Since the beginning of what has been called the "Ritalin wars" that began in the early 1990s, this has been one of the most contested aspects of the ADHD controversy in the United States: that ADHD is primarily a genetically based (neuro-biological) disorder having less to do with environment and sociopolitical inputs such as home life, community health level, or high-stakes schooling contexts, and more to do with heritability.[9] Lawrence Diller, a physician who has spent a great deal of his career studying the permutations of the ADHD controversy in the United States, provides two important factors that a shift from a psychological basis for ADHD to a psycho-biological allows drug companies in terms of controlling the knowledge parameters for understanding ADHD as a disease. First, Diller (1998) states, "the power of brain-based explanations of ADD for people whose lives are profoundly stressed by their child's behavioral symptoms, or their own . . . they [parents and teachers] badly want answers, and most are frustrated by suggestions that their lives may be out of balance, or their kids' schools are understaffed, or their responses to a child's acting out are ineffective. It's much simpler to call it a neurological problem" (103). Here, a market need is not only met by pharmaceutical companies, but by framing ADHD as a genetic and biological problem they become the only viable answer for a cure. Second, Diller (1998) also points out that the shift from psychological to genetic grounds in terms of how teachers, school personnel, and parents think about ADHD "implicitly diminish[es] the significance of learning disabilities and emotional problems, family dynamics, classroom size, and economic and cultural issues that may be relevant to ADD, in favor of genetic and neurochemical factors" (103).

One of the chief ways the biological paradigm has become the governing rationale in schools for diagnosing and treating ADHD has been through the work of the NIMH. As the governmental authority on mental health issues (and one of the 27 divisions of the National Institutes of Health [NIH]), NIMH's mission "is to transform the understanding and treatment of mental illness through basic and clinical research, paving the way for prevention, recovery, and cure" (NIMH 2012). NIMH is also the leading funding source for research on mental health in the United States. Not surprisingly, the NIMH (and its parent agency the NIH) has been embroiled in controversy surrounding its flagrant ties to the pharmaceutical industry, most notably as a publically funded government agency that pays for private research and development in universities by some of the biggest drug companies in the world. Most recently, the director of NIMH, Thomas Insel, has been called before a House subcommittee investigating ties between the NIMH and the pharmaceutical industry. The smoking gun under investigation by the House subcommittee was Insel's ties to Dr. Charles B. Nemeroff, a long time drug researcher who, for over a decade, accepted large sums of money directly from the pharmaceutical industry while working on NIMH grant funded research. Nemeroff ended up leaving Emory University after "U.S. senate investigators found he received $2.8 million from GlaxoSmithKline and other pharmaceutical companies between 2000 and 2007 and failed to disclose at least $1.2 million of it" and landed a new position at the University of Miami where Nemeroff has been allowed and even solicited to apply for further NIMH grant funding (Basken 2010).[10]

The story of Nemeroff is not one of a lone gunman. Rather, it is indicative of the neoliberal research and development culture that industry and universities have fostered since the Bayh–Dole Act made corporate and academic research indistinguishable endeavors. As Marcia Angell (2004) discloses in her insider look into the co-productive relationship between the pharmaceutical industry and NIH-funded scientists, "big pharma has compromised the research community [through] its extensive inroads into the NIH itself" (104). This has largely occurred by "senior NIH scientists (who are among the highest paid employees in the government) routinely supplement[ing] their income by accepting large consulting fees and stock options from drug companies that have dealings with the institutes" (Angell 2004, 104).[11] The reason such an unsavory history is necessary to recount is that, as the leading institution and knowledge industry controlling and disseminating how mental illness and treatments are to be understood in society (and as we will see in schools), the NIMH is the most influential actor involved in the pharmaceutical industry/parent/school network. As such, the research pertaining to ADHD that is attempting to strengthen the case for the biological origins of the disorder must be viewed within the light of this co-productive relationship between biocapitalism and higher education institutions.

The NIMH has also been one of the leading advocates for using drug therapy strategies as the primary approach for dealing with ADHD children. Or, as Peter Breggin (2001) has put it, the "NIMH has vigorously promoted the interests of giant corporations that peddle psychiatric drugs to people . . . Many of the most ardent drug advocates in this nation are past or present full time employees of NIMH. Most of the others regularly benefit from large NIMH grants and other government perks. Much of the flawed research purporting to show a biological basis for ADHD has been carried out or funded by the NIMH" (141). In the most famous NIHM push for standardizing psychotropic drug intervention as the most effective treatment for ADHD children, a NIMH-funded (in collaboration with the DOE) study was published that unequivocally promoted drug therapy as the optimal treatment for school children afflicted with ADHD. In the study's conclusion, for example, the authors asserted that "for ADHD symptoms, our carefully crafted medication management was superior to behavioral treatment and to routine community care that included medication" (The MTA Cooperative Group 1999, 1073).[12]

The construction of a public understanding (especially targeting teachers) of ADHD on biological grounds such as the one the NIMH promotes has a number of important implications to the formation of biological citizenship within school settings. From the standpoint of teachers, a normalized understanding of ADHD as a biological abnormality in students has begun to create a situation where, as Nikolas Rose (2007) argues in his research on the formation of biological citizenship, "biological and biomedical languages are beginning to make up citizens in new ways in the deliberations, calculations, and strategies of experts and authorities: for example the emergence of categories such as the child with attention deficit hyperactivity disorder, the woman with premenstrual dysphoric disorder, or the person who is presymptomatically ill because of genetic susceptibilities" (140). Within schools, teachers are playing (and have been for quite some time) a key role in defining the biological citizenship of students: not only in those students who fit the ADHD profile and are recommended for diagnosis, but also by generally understanding student populations as genetically deficient or as non-diseased groups.

NIMH's most recent attempt to anchor ADHD in a biological framework and to maintain the position that psychotropic intervention is the best form of treatment involved the utilization of brain imaging technology on afflicted children. While there have been multiple NIMH studies of ADHD using different brain scanning technologies (such as PET scans), the most recent study used magnetic resonance imaging (MRI) to measure frontal lobe cortex thickness of ADHD and non-diagnosed children.[13] What is new this time around in the latest NIMH study, however, is not only the use of MRI technology to render empirical evidence that suggests "children with ADHD show relative cortical thinning in regions important for attentional control," but also that from such images of children's brains, a more accurate and appropriate treatment plan can be established (Shaw et al. 2006, 540). The NIMH's latest position on ADHD moves further toward a biological or genetic explanation for ADHD that has at least two important implications that affect the biopolitical terrain of neoliberal schooling.

First, one of the major findings of the NIMH-funded MIR study of ADHD children is that clinical practices may soon lead to tailored pharmaceutical drug treatment plans designed for specific individuals' ailments (or in this case what brain imaging suggests). As James Kennedy, a prominent professor of psychiatry, puts it, "with more research, it may be possible to do an MRI study before starting medication and then predict what type of treatment might be best for that individual based on their brain image and genotype" (Singer 2007). The tailoring of psychotropic drug combinations based on individualized biological information (biometrics) opens the door for the creation of more ADHD drug treatment plans and, in turn, a greater market need for a wider variety of psychotropic drugs. The second important implication of the NIMH's brain imaging study is that the official (governmental) stance on ADHD has become decidedly more rooted in a biological and genetic explanation. As a result, schools and governmental and corporate agencies that shape ADHD diagnosis and treatment protocols for children are even more wedded to a genetic model of intervention. Moreover, teachers (the primary source of almost all ADHD diagnosis referrals) are subjected to a unitary vision and understanding of ADHD that helps construct teacher strategies for dealing with behavioral issues in the classroom. In other words, the formation of biological citizenship in the school is heavily influenced by a pedagogical regime that comprises the NIMH and other official actors that premise ADHD on genetic grounds. As such, student behavior matching ADHD symptoms are immediately targeted as a biological problem requiring a neurochemical solution. Finding this model of biological citizenship at work in the school is not difficult. In fact, it is the one promoted by the DOE's ADHD manual for teachers and the leading educational psychologists' guideline book for ADHD intervention.

The Biochemical Governing of Bodies in School

At this point, it will be helpful to provide the American Psychiatric Association's (APA) clinical definition of ADHD. The APA's definition is important to consider because it frames the DOE's educational manual for teachers, parents, and school personnel designed to help identify and treat ADHD in children. Perhaps more importantly though, the APA's clinical definition of ADHD represents a field of knowledge that helps construct a particular regime of practices in schools used for identifying and treating ADHD students. From a governmentality perspective, understanding how these two forms of power operate and inform each other (expert knowledge of a disease and practices of disciplining and controlling the disease in social spaces) helps us understand more concretely how a type of biological citizenship around ADHD comes into being within schools. What I want to accentuate in other words is how ADHD in schools is connected to "regimes of practices [that] give form to and are informed and reshaped by various forms of knowledge and expertise," and how the disorder is enmeshed in a system that wants to "codify appropriate ways of dealing with them [i.e. questions of health and education], set the aims and objectives of practice, and define the professional and institutional locus of authoritative agents of expertise" (Dean 2010, 32). The DOE's teacher's manual and leading guide for teacher and school personnel on ADHD reflect precisely the type of governmentality associated with ADHD in schools.

The DOE's publication *Identifying and Treating Attention Deficit Hyperactivity Disorder: A Resource for School and Home* (2003) presents teachers the following definition of ADHD that comes directly out of the APA *Diagnostic Statistical Manual of Mental Disorder*:

Inattention

1. often fails to give close attention to details or makes careless mistakes in school work, work, or other activities;
2. often has difficulty sustaining attention in tasks or play activities;
3. often does not seem to listen when spoken to directly;
4. often does not follow through on instructions and fails to finish schoolwork, chores, or duties in the workplace (not due to oppositional behavior or failure to understand instructions);
5. often has difficulty organizing tasks and activities;
6. often avoids, dislikes, or is reluctant to engage in tasks that require sustained mental effort (such as schoolwork or homework);
7. often loses things necessary for tasks or activities (e.g., toys, school assignments, pencils, books, or tools);
8. is often easily distracted by extraneous stimuli.

- Hyperactivity

1. often fidgets with hands or feet, or squirms in seat;
2. often leaves seat in classroom or in other situations in which remaining seated is expected;

3. often runs about or climbs excessively in situations in which it is inappropriate (in adolescents or adults, may be limited to subjective feelings or restlessness);
4. often has difficulty playing or engaging in leisure activities quietly;
5. is often "on the go" or often acts as if "driven by a motor";
6. often talks excessively;
7. often blurts out answers before questions have been completed;
8. often has difficulty awaiting turn;
9. often interrupts or intrudes on others (e.g., butts into conversations or games).

The DOE's teacher's manual goes on to state that children meet the clinical definition of ADHD if they exhibit six or more of the above symptoms in either of the two main categories, inattention or hyperactivity (DOE 2003). While I will not discuss the wide parameters established for the disorder in the APA's diagnostic metric for ADHD, it should be noted that classroom contexts increasingly driven by stricter accountability measures and canned curriculum certainly seem to produce the conditions where many students might exhibit any number of the listed "symptoms". What I am most concerned with instead is the DOE's pedagogical project embedded in the ADHD teacher manual. It is one that promotes a model of biological control that has strongly contributed to the shaping of school approaches to behavioral intervention toward students. In particular, what teachers and families learn from the DOE's diagnostic manual is (1) that ADHD is a genetic problem most effectively managed through psychotropic drug intervention and (2) school practices for governing abnormal bodies (those diagnosed with ADHD) should enact the rationality for the biological control established by institutions such as the NIMH and DOE. Let us look at each of these factors in turn.

As primarily a pedagogical tool for teaching educators and parents how to identify ADHD in children, one of the key lessons drawn by the DOE's teacher's manual is that the biological view of ADHD is presented as objectively certain; a genetic interpretation is the natural way in which to understand the condition which is explainable through new advancements in scanning technologies such as MRI's of the brain. For example, on the first page of the manual, readers are told "we are now learning that ADHD is not a disorder of attention, as had long been assumed. Rather, it is a function of developmental failure in the brain circuitry that monitors inhibition and self-control" (DOE 2003, 1). Under the section "What Causes ADHD?" the manual goes on to assert that ADHD is almost entirely a biological problem: "Most researchers suspect that the cause of ADHD is genetic or biological, although they acknowledge that the child's environment helps determine specific behaviors. Imaging studies conducted during the past decade have indicated which brain regions may malfunction in patients with ADHD, and thus account for symptoms of the condition" (DOE 2003, 2). The pedagogical tone of the entire manual is substantiated with scientific studies and clinical research (most prominently from studies done by the NIMH) on ADHD which all but bracket out any competing explanation for the condition. In fact, the ADHD researcher most cited throughout the teacher's manual is Russell Barkley, one of the most outspoken advocates and author of multiple popular books and research articles promoting a genetic explanation and aggressive drug treatment intervention approach for ADHD children and adults.[14]

The DOE teacher's manual, however, is much more than a suggestion guide for teachers and parents who want to learn more about ADHD and how to spot the disease in children. There is also a deeper pedagogical literacy being promoted that portrays children's bodies through the language and cartography of neurochemical pathways, genetically diseased groups, and ultimately biochemical machines that can be manipulated and adjusted for optimal performance. What I am suggesting lies beyond the controversy between those in the psychiatric field and public who are against a biologically or genetic dominant explanation for ADHD and those who promote it is a particular discursive knowledge field that opens the child's body up to a variety of new forms of biological control. As I pointed out in the earlier section, it is a discursive field of knowledge that teachers, school personnel, and administrators have adopted in their practices for dealing with students presenting ADHD symptoms. It is also a type of rationality that opens up the body to a deeper field of technological control that does not function necessarily on the surface of the body as did the spanking paddle, but rather intervenes into neuro-pathways that one's genetics did not get exactly right. In educating a biological/genetic literacy used to frame understandings of ADHD, the DOE's teacher's manual constructs a crucial component of the type of biological citizenship that surrounds ADHD diagnosis and treatment, and specifically the role schools play in the process. Yet the biological/genetic literacy that underpins the DOE's teacher's manual is only one part of the equation of how biological citizenship is being shaped inside schools. There are also the concomitant practices of regulating children's bodies through an ever-expanding array of biochemical tools that comprise the model of ADHD governmentality that has emerged in schools in the age of biocapitalism.

As I have pointed out earlier, one of the key parts to the emerging model of biological citizenship as it relates to ADHD in schools is how it normalizes the disorder as a biological problem. The second feature of biological citizenship in schools I am concerned with is how the biological/genetic rationality taught through the DOE manual as a regime of knowledge invokes a type of governmentality over children's bodies. The model of governmentality I am suggesting is embodied in the DOE manual can perhaps best be set into relief by posing this question: How does the dominant expression of biological citizenship that is shaped by genetic explanations of ADHD call for certain practices or action on the student's body? And second, how do the practices that are hailed into the everyday life of schools inform understandings of self both by individuals who are the target of these technologies of control and by the people who carry through the governing of children's bodies? Again, the DOE's teacher's manual gives us a mooring point in which to view how the biochemical governing of ADHD children is constructed in schools. In the section that seeks to answer the question "what are the treatment options?" the manual quickly dismisses the effectiveness of behavioral therapy as a reliable treatment approach. The manual states that "the research on the effectiveness of behavioral techniques are mixed. While studies that compare the behavior of children during periods on and off behavior therapy demonstrate the effectiveness of behavior therapy, it is difficult to isolate its effectiveness. The multiplicity of intervention and outcome measures makes careful analysis of the effects of behavior therapy alone, or in association with medications, very difficult" (DOE 2003, 10). Here, by questioning the effectiveness of behavioral therapy, it is strongly suggested to teachers and families that behavioral therapy is simply not an efficient and measurable treatment strategy for ADHD children. At best, the DOE manual asserts, behavioral therapy is most effective when used in combination with psychotropic drug therapy (DOE 2003).

In adopting the biological/genetic model wholeheartedly, it is only logical that in the DOE manual behavioral therapy is framed as ineffective and not accurately measurable when compared to psychotropic intervention. Indeed, if, as one of the primary researchers cited in the DOE manual believes, ADHD is as much as 80 percent genetic (only 10 percent less than what scientists believe determines an individual's height), then how could behavioral therapy possibly be effective in re-routing the neural pathways of ADHD children? Behavioral therapy options become even more irrelevant within the biological landscape of the DOE ADHD manual, given that pharmacological approaches are asserted to achieve a 75–90 percent effective rate in children with ADHD (DOE 2003). In essence, the DOE manual employs a discourse of the human body that calls into reality practices born from a specific field of knowledge: genetic science. For instance, if a biochemical blueprint of students' bodies are used for understanding how to effectively manage the problem of ADHD in schools, then the following proscription for practice is a natural outcome: "Researchers believe that psychostimulants affect the portion of the brain that is responsible for producing neurotransmitters. Neurotransmitters are chemical agents at nerve endings that help electrical impulses travel among nerve cells. Neurotransmitters are responsible for helping people attend to important aspects of their environment. The appropriate medication stimulates these underfunctioning chemicals to produce extra neurotransmitters, thus increasing the child's capacity to pay attention, control impulses, and reduce hyperactivity" (DOE 2003, 11). Here, the shift from knowledge about the body (neurotransmitter deficiency as a product of genetic determination) leads logically to adjusting the body biochemically (underfunctioning chemicals). Discourse and practice form an elegant loop in which governance of the child's body in school is thought about and carried through with technologies that allow for biochemical control.

From a governmentality perspective, one interested in understanding how scientific discourses articulate through institutional channels of power (the DOE and NIMH, for instance) and shape fields of action where subjects are regulated and controlled through corresponding sets of practices (school ADHD intervention strategies), we must also see how biochemical governing strategies associated with ADHD co-articulate with human capital frameworks of education. The biochemical governing of children in schools, stated differently, also needs to be understood within the larger network of power relations regulating educational life in ways that maximize the biocapital value of students in the neoliberal era. That is, one way to read the ADHD definition and correlating symptoms presented earlier is as a character list for a hobbled human capital subject. Inattention and hyperactivity, the broadest general categories of behavior associated with ADHD, are diametrically opposed to habits and behaviors generally thought of as leading to productive human capital accumulation in students such as good test taking skills, ability to concentrate for long periods of time, and conforming to highly structured and standardized learning environments. Thus, one of the major underlying goals of biochemical governing is to get "the child's behavior and disposition . . . to a point where the child can perform optimally in both academic and social settings, while keeping side effects at a minimum" (DOE 2003, 11). This coupling of neoliberal educational imperatives (the cultivation and production of a healthy and useful human capital stock) and forms of biochemical governance is one of the most alarming trends emerging from the coalescing of education and biocapitalism. Indeed in the co-articulation of biochemical governance that underlies the formation of biological citizenship as it pertains to ADHD in schools with the human capital imperative to optimize the value of the educational population, one of the more insidious neoliberal strategies for extracting more value from educational life is apparent. The overlapping goals of ADHD management through biochemical governance and increasing the quality of educational biocapital are perhaps even more apparent in another leading practitioner's guide to ADHD in the school.

George DuPaul and Gary Stoner's (2003) bestselling reference and text *ADHD in the Schools: Assessment and Intervention Strategies* is designed to "meet the needs of a variety of school-based practitioners, including school psychologists, guidance counselors, administrators and both general and special education teachers" (xiv). As the widely accepted ADHD bible for school practitioners, DuPaul and Stoner's text offers a mainstream understanding of how psychotropic therapy intersects with student efficiency imperatives that define neoliberal educational goals. DuPaul and Stoner's book thus provides practitioners with a guide to the vast evidence-based literature on ADHD strategies for intervention, and as such exemplifies the most important and neglected sub-discourses that help constitute biochemical governance in school settings: a focus on how individual students' academic performance is positively altered through drug intervention. In other words, part of making ADHD students "normal" in DuPaul and Stoner's text includes measuring student performance optimization before and after students' neurological pathways have been re-routed through psychotropic drugs intervention. DuPaul and Stoner (2003) invoke the performance imperative associated with drug therapy intervention saying, "research studies conducted by several independent research teams have found acute MPH-induced [methylphenite which is the active amphetamine substance found in Concerta, Ritalin, and partially in Adderall] in academic productivity and accuracy among large samples of children and young adolescents with ADHD. Attention to teacher lectures, completion of study hall assignments, and quiz and test scores among junior high school students with ADHD also are enhanced by MPH" (197). Moreover, DuPaul and Stoner (2003) assert that there is strong empirical evidence to support that students, when looked at the group level, perform best in the classroom at a higher dose range (197). The unit of measure being studied here between classroom performance and dosage rates of psychotropic drugs is revealing as to how biochemical governance understands and disciplines the body. ADHD children become "normal" depending on how well biochemical alteration through drugs increases students' measurable forms of educational achievement such as standardized test scores and teacher observation comments that are part of the research methodology of many ADHD test studies. Here, mental disorder and academic performance are conflated into one and the same problem: how to discipline and regulate student bodies so as to enhance classroom functioning and thus the overall surplus value of human capital in the classroom.

DuPaul and Stoner's analysis of the "normalization of classroom functioning" through psychotropic drugs is also telling as to how the effects of biochemical governing are measured in classroom settings. Citing a study that examined the "degree to which MPH normalized the classroom behavior and academic functioning of 76 children with ADHD" against a normal control group of 25 children, DuPaul and Stoner praise the correlation between biochemical governance and increased academic performance. DuPaul and Stoner (2003) submit:

> Children with ADHD participated in a double-blind, placebo-controlled trial in which they received each of four doses (5, 10, 15, and 20 mg) of MPH and a placebo. Dependent measures included teacher ratings of social conduct, direct observations of classroom on-task behavior, and accuracy on independent academic tasks. Between 53% and 78% of the sample obtained scores within the normal range of functioning at one or more doses of MPH depending upon the specific measure employed. Notably, teacher ratings of behavior control showed the most prominent improvements and normalization, followed by direct observations and task-related attention and academic efficiency. (201)

There are two dynamics pertaining to biochemical governance that I want to highlight in the above study offered by DuPaul and Stoner. First, I want to clarify what exactly the study is attempting to measure: the neurochemical effect of psychotropic drugs on student performance in classroom environments. What cannot be lost from sight is the fact that such a study is related to biological/genetic knowledge that understands ADHD as a deficient configuration in the body's neural circuitry. Put differently, what is being tested is not only how psychotropic drugs effect student performance, but also the very genetic basis of the disorder itself. If results suggest, in other words, that drugs are having a positive effect on ADHD children's academic performance, then by extension ADHD is a biological/genetic problem to be handled most effectively through biochemical means. Again the sub-discourse here points to a particular model of biological citizenship and its corresponding model of bodily governance.

The second dynamic at play in the study offered by DuPaul and Stoner comes together in the role of the teacher as one of the most important indicators of student performance. On the one hand, this makes complete sense. Teachers are seemingly the best people to be able to observe and determine the behavioral effects psychotropic drugs have on their students. On the other hand, the study also situates the teacher within a co-productive framework to the study itself that is interested in determining the relation between biochemical intervention and student performance. Teachers in the study participated in making knowledge about ADHD governing; that is, in attempting to draw linkages between psychotropic drugs and the "normalization of classroom functioning," the teacher also became a knowledge coproducer about the disorder. This knowledge coproducing role taken on by the teacher in the study is important on at least two important levels. At the level of practice, education in such a study is defined as something the teacher can achieve only through a process of normalizing students to accepted classroom behavior such as ability to stay on task, test well, follow instructions, and so on. The way we can know this definition of education is operating in the study is how ADHD students are understood to be abnormal, not able to function optimally in the classroom. Thus, the practice of education presented in the study used by DuPaul and Stoner is about developing bodily practices that can maximize the normalcy of educational life in the classroom.

The second level I want to draw attention to in the study is the overall pedagogical message being sent to teachers, school counselors, psychiatrists, and administrators. As one of the most utilized texts for educating teachers and school personnel on ADHD intervention strategies, situating teachers as coproducers of knowledge about ADHD best practices for drug intervention naturalizes the view that teachers are important gatekeepers to what is considered normal and abnormal, efficient and inefficient in the classroom. It places teachers in a pseudo-clinical position where part of good teaching is blurred with an ability to govern students' bodies through a variety of technologies of control, including biochemical. The marrying of teaching to ADHD diagnostic practices unites genetic understandings of the body with strategies for controlling it. In short, the acts of teaching and biochemical governance converge in studies such as DuPaul and Stoner's where maximizing student performance through the disciplining and measuring of the educational body is portrayed as just a part of everyday classroom management.

Alternative Brave New Worlds

As I have argued in the preceding sections, there is much more to the ADHD epidemic controversy than skyrocketing diagnosis rates and prescription drug usage by children. While these statistics are troubling enough, especially considering that almost 10 percent of the K-12 educational population in the United States has been diagnosed with some form of ADHD, my focus in this analysis points to an equally alarming model of governmentality that has coevolved with the ADHD crisis. The dominant governing strategy associated with ADHD regulation in schools that I have called biochemical governing is one that has emerged from a confluence of technoscientific knowledge that situates the disorder as a genetic problem with best practices oriented toward managing sickness that effects learning through psychotropic intervention. What is more, in the neoliberal era of education where human capital optimization is paramount, the need for technologies and practices that can squeeze the most value out of educational life is only becoming greater. The ADHD crisis and high-stakes schooling thus represent a dangerous pairing: a genetic problem that hinders human capital accumulation (one of the primary goals of neoliberal education) and a perceived societal need to maximize the latent value of both students *and* teachers. Within such a context, biochemical governance, I argue, should not be thought of as an anomaly or an extreme attempt to manipulate students' bodies for optimal performance in the classroom. Rather, biochemical governance within our neoliberal educational moment should be understood as a part of a continuum of control strategies oriented toward the development of a quality human capital stock within a globally competitive market society. Special education programs used as holding tanks for "problem" students, mechanisms in schools that contribute to high dropout rates of students "who don't want to be in school anyways," value-added metrics coupled with merit pay systems, competitive enrollment policies for high performance charter schools, or simply urban schools designed with high levels of bodily surveillance and military-like curriculum structures are all biopolitical tools used and designed for sorting and optimizing the productive energies of the educational population in the United States. Each of these technologies of control constitutes the biopolitical constellation shaping education reform taking place under neoliberal restructuring policies in the United States such as Race to the Top.

What is unique about biochemical governance associated with the ADHD epidemic in this country, however, is how genetic knowledge of the body and biotechnological control of student behavior are applied to increase human capital value among individual students and groups. In this sense, biochemical governing can be seen as the most acute example of a biopolitics of education. Educational life, the actual body, and its organic systems are the target of technologies of control in a very real way. Yet, as I have also argued, one of the most important reasons biochemical governance has become the hegemonic manner in which to regulate student bodies identified as presenting ADHD symptoms in schools has to do with the normalized model of biological citizenship associated with ADHD that allows and even calls for psychotropic solutions to genetic problems. If deficient neurological pathways are the way teachers, families, and school personnel are taught to understand and act upon ADHD bodies, then there is very little room for alternative ways to mediate behavior issues associated with ADHD symptoms in students. What I am proposing such a situation calls for is the development of an alternative model of biological citizenship, one not based on the power and ethical relations that discipline the body in the way that biochemical governance does.

Part of the challenge of creating an alternative model and practice of biological citizenship, however, is the school itself. What I mean by this is schools in the United States have always functioned as technologies of control that have shaped the US population in line with dominant governing rationalities of the state and its productive needs from the population, be it for labor, security, racial purity, or projects of territorialization. American Indian boarding schools and legally segregated schooling were two of the clearest examples of how the public school system in the United States socially enframed ways of thinking about "degenerate" or "savage" segments of the population and how the state could best manage these problems within its population for economic and racial advantages enjoyed by whites and industrial leaders. The models of biological citizenship operating in these examples were ones shaped by eugenic science and white supremacist notions of how to govern the people of a nation through a racial contract and colonial project (Willinsky 1998; Mills 1999). Schools served a very important function within these projects. The type of governance I have been mapping in this chapter does not necessarily operate exactly the same way as these two historical examples, but certainly belongs to the same sovereign family. When problems such as behavioral and disposition abnormalities in children are dealt with in an existing technology of control such as the school, especially schools that are undergoing neoliberal structural transformations of a radical kind, the outcome has been the development of a type of biological citizenship in schools that allows students and teachers to be acted upon as deficient and suboptimal bodies not capable of producing adequate surpluses of human capital. As such, I think the best places to begin looking for alternative models and practices of biological citizenship are spaces outside schooling structures that can help us reimagine what education may look like inside the school. Let me offer an example.

Alternative models and practices of biological citizenship should start from biosocial communities experimenting with and practicing a different somatic ethic—the ethical relations we have learned to apply to our corporeal bodies—rather than those embodied in my analysis of biochemical governing associated with ADHD currently promoted in neoliberal school settings. Biosocial communities are, as Paul Rabinow (1992) and Nikolas Rose (2007) have identified in their research, groups and communities of people who pragmatically negotiate, revise, fashion, and manage understandings of the self—specifically how individuals conduct themselves in relation to biomedical and biotechnological regimes of knowledge and practice in society. Where would one look for a biosocial community that might offer a counterexample from which to learn?

Even in the example of ADHD biological citizenship I have traced in this chapter, there is room for different power relations to emerge—an alternative biosocial community to the one currently subject to forms of biochemical governances and the power relations that create such a regime. In other words, biosocial communities are not closed circuits or predetermined spaces with impermeable boundaries. Rather biosocial communities are contested, shifting, and constructed in relation to evolving discursive and practical regimes of the body. Many ADHD parents, for example, have created alternative approaches and forms of resistance against dominant diagnosis and treatment plans associated with ADHD, such as adjustments in diet, holistic medicine, or removing students from stressful, high-stakes educational environments altogether (though most of these are counterstrategies that require a privileged socioeconomic status). Yet, as I have also argued earlier, the setting of the school, especially under severe neoliberal restructuring policies, is a particularly barren place to experiment with alternative biosocial communities that are resistive to governmentalities ultimately rooted in rationalities and practices interested in optimizing the bodies of students and teachers. I suggest instead that alternative biosocial communities need to be identified where biological citizenship is constituted through resistance to the further enclosure of biological life that neoliberal rationalities and practices have escalated. One such example of a biosocial community can be found in the alter-globalization movement in Southern Mexico that has been resisting and struggling against transnational agribusiness and, in particular, the genetic preservation of maize (corn) and place-based farming and cultural practices.

The reason I am suggesting that the ADHD epidemic and genetic pollution of indigenous corn in Southern Mexico caused by neoliberal development policies such as the North American Free Traded Agreement (NAFTA) are similar in some ways is because they both represent an example of biological enclosure. In other words, both NAFTA's farming policies that mandate GMO corn growth and industrial agricultural practices and ADHD biochemical governance represent a further privatization of life and how public understandings of life are increasingly being transformed through emerging bioscientific and biotechnological discourses and practices.[15] In this sense, what farmers and communities members have been doing in their struggle against transgenetic drift from GMO corn into the biodiverse rich corn areas of Oaxaca and Chiapas since NAFTA was institutionalized in the mid-1990s, is protecting *both* the life of indigenous corn and the cultural practices and knowledge that are connected to the plant from enclosure. What has also developed in the resistance movement against transgenetic pollution of corn and culture in Southern Mexico is a type of biological citizenship that is rooted in food sovereignty politics—where a politics of protecting and preserving cultural food practices creates an ethical and health relation to the body autonomous to neoliberal strategies of development and biocapitalist expansion. Farmers and community members' ongoing struggle against biocapitalist growth in Southern Mexico, in other words, offers an alternative type of biological citizenship to learn from that rejects the genetic control of the body (and plants) instantiated through neoliberal development policies driven by big agribusiness and transnational trade agreements.[16]

One important reason the combined food sovereignty and autonomous community movements in Southern Mexico are helpful for rethinking the model of biological citizenship associated with ADHD is because they are producing a biosocial community built on a qualitatively different somatic ethic—a community and individual political recognition that life is at stake in multiple senses. In her work that has looked at how indigenous farming communities have organized and developed a politics of resistance against the "genetic pollution" caused by neoliberal farming policies in Mexico, Kathleen McAfee

(2003) describes what I am suggesting are important features of an alternative biosocial community to learn from:

> Through farmer-to-farmer exchanges, they develop and apply principles of agroecology, using traditional and new methods, low inputs of external energy and chemicals, and intensive reliance on site-specific farmer intelligence. For them, agroecology is more than a technological means of conserving biodiversity and increasing productivity, although some have done so with impressive success. It is also a means toward greater social equity and local control over food sources and supplies, and the core of a social and environmental alternative to neoliberalism. (35)

In biosocial communities such as those McAfee describes, the development of biological citizenship is rooted to an ethic of the commons. That is, at stake for biosocial communities such as those in Southern Mexico is the enclosure of not only land but also the genetic biodiversity of maize and community practices and knowledge of agroecology that have developed over time in a specific socio-ecological context. In regions such as Southern Mexico or for that matter many other communities across the globe organizing resistance movements around genetic encroachment into more and more areas of life, forms of neoliberal governmentality strategies have become a focal point of politics and for rethinking what should (or can) be put in our bodies in terms of food and who should be making these decisions. What I am suggesting can be learned from biosocial communities engaged in resistance to technoscientific food products and neoliberal developmental policies that promote their proliferation is in the way such political models center the question of how forms of life are being enclosed and, in turn, experimented upon toward the creation of alternatives to forms of enclosure.

Just as the communities in Southern Mexico have recognized and organized in their anti-globalization movement, once new zones of life (neurological circuits) become part of the normalized discourses and practices of neoliberal governmentality strategies then the very biochemical terrain of our bodies are also subject to a kind of enclosure. The biomedicalization of students' bodies, in other words, can also be seen as a territorialization of neurological pathways, biochemical communication, and, perhaps most importantly, how individuals and groups think about their bodies and ways of altering their perceived "deficiencies." Here again we see the neoliberal desire to regulate and manage life in the field and the classroom united into a disturbing symmetry. It is this merging between governmentalities of life and neoliberal educational reform strategies that must be challenged through inventive and practical means. One strategy of resistance that farmers in Southern Mexico have developed is to network with educational groups that support and promote autonomous models of education such as the Schools for Chiapas program.

Created in the mid-1990s, Schools for Chiapas is a grassroots program that began by raising funds to support the constructions of schools in the autonomous Mayan communities of Chiapas and is based out of San Diego, California. The autonomous schools that have grown in number over the years in Southern

Mexico have developed curriculum rooted in the ethic of "protecting the health and culture of the Mayan communities by educating for sustainable, organic, and dignified farming methods" (Schools for Chiapas 2012). One of the centerpieces of the autonomous schools' curriculum is the promotion of agroecology and the Mother Seeds in Resistance project "which is designed to eliminate transgenetic contaminations of native landraces of corn in Chiapas, Mexico" (Schools for Chiapas 2012). Built into the practice of education in the autonomous schools, in other words, is a learning model that centers on resisting the enclosure of life through genetic pollution that includes testing corn plants for genetic cross-pollination, the establishment of long-term seed storage facilities, and networking resistance projects with educators and communities across the globe by providing indigenous corn seeds as part of a "solidarity grow-out." The Mother Seeds in Resistance effort, one pillar of the Schools for Chiapas educational philosophy and practice, has three projects it lists as part of the autonomous education program: (1) "Long term preservation under community control of natural corn seed," (2) "statewide genetic testing of corn by Zapatista ecological-agricultural promoters," and (3) "solidarity growing of Zapatista corn by supporters everywhere" (Schools for Chiapas 2012).

What I am suggesting autonomous educational networks such as Schools for Chiapas represents is a real model in which to experiment toward alternative conceptualizations and practices of biological citizenship within educational settings. In particular, the Schools for Chiapas autonomous schooling network provides a model of an affirmative biopolitics of education, one where students, teachers, and community members produce social life in resistance to neoliberal forms of governmentality that threaten health, socio-ecological relations, the right to use and develop sustainable tools and practices that provide the means for food sovereignty, and a community-networked education that is critical of the ways the local and global are being mobilized for biocapitalist needs and territorialization projects. Instead of biochemical governing and the unhealthy biopolitics it promotes in schools, new seeds need to be planted that can grow into alternative futures. As my analysis of biochemical governing associated with ADHD has demonstrated, life itself has become the terrain in which resistance and real alternatives must be created. Not unlike GE monocropping, biochemical governing is an act of enclosure that is quickly becoming a routine part of schooling in the biocapitalist era. In a historical moment when the neurological circuitry of children's brains and the genetic structure of corn have both become part of the neoliberal calculus of efficiency and optimization practices of resistance should start with untangling vitality from regimes that produce something like biochemical governance. One place to start may be to plant non-GMO corn from Chiapan farmers in a learning garden as a starting point for examining the interrelated networks of biocapital and biosocial communities. Here how the corn is controlled through technoscientific means can lead to critical discussions and analysis of other forms of biotechnological control over life such as the type of monocropping biochemical governmentality is attempting to achieve in the classroom. Returning to the seeds of life may be a good point of departure for redefining life outside of biocapitalist expansion.

Epilogue

> The ability, however, not only to live in new ways, but to insist on this freedom demands that we clearly recognize what distinguishes the perception of *homo economicus* from all other human beings.
>
> —Ivan Illich

Alternative Futures of Education: Exiting Education for Biocapital

This book has been an attempt to understand and map the colliding enterprises of education and biocapitalism. Biocapitalism in its broadest sense, as I have tried to capture the aspects of in each chapter, is a model of political economy that has further opened earth's biology (both human and nonhuman) through technoscientific advances to the dynamics of a rapidly expanding free-market capitalism. Without the discovery of rDNA and the largest industrial agricultural system in the world, for example, genetically engineered food would not be one of the fastest growing areas of the industrial food economy. If education in the United States has been enlisted in the project of biocapitalist development (both as a productive and a consumptive population) where do we begin to look for educational alternatives? In other words, in learning contexts that increasingly look like the GMO cornfield that is managed at the molecular level, where do we start identifying both practical and theoretical points of departures from biocapitalist modes of education?

One of the often overlooked aspects of the biopolitical tradition is its "affirmative" formulation. That is, in addition to the focus on the ways life has become increasingly subjected to regimes of neoliberal governmentalization, biopolitics encompasses not only "how power works for and through subjects but also at the potential for the production of alternative subjectivities" (Hardt and Negri 2009, 59). For example, Thomas Lemke (2011) points to such an "affirmative biopolitics" in the work of Robert Esposito in which a "main point of reference is to incomplete and open individual and collective bodies"; moreover, "these bodies defend themselves against attempts at identification, unification, and closure and articulate an immanent normativity of life that opposes the external domination of life processes" (91). In the field of educational biopolitics, Henry Giroux (2008) has also called for a type of "affirmative biopolitics" where it "is also potentially about enhancing life by linking hope and a new vision to the struggle for reclaiming the social, providing a language capable of translating individual issues into public considerations, and recognizing that in the age of new media the terrain of culture is one of the most basic precepts of the new authoritarianism" (61). Tyson Lewis, in his work, has signaled the need for developing a type of biopedagogy in response to the negative forms of biopower that organize and regulate student's lives in educational spaces under Empire. For him, a biopedagogy is one that can create a "generative life through which new notions of democratic community can evolve" (Lewis 2007, 700).[1]

What is missing, however, and what I have tried to articulate through different registers in this book, is a map of (1) where and how biocapitalism and education are most clearly in a state of co-production and (2) possible starting places to develop educational practices and modes of being that resist, reject, and refuse how life is understood and treated as an investible and extractive value within learning spaces shaped by biocapitalism. Such a cartography of the merging of biocapitalism and education is important I believe because if an alternative biopolitics of education is going to be developed, then we need to be clear how existing power relations exercise forms of control and regulation within educational zones. For me this requires a simultaneous analysis of biocapitalism and education since I see the two as co-productive partners that are shaping the current historical moment in some very profound ways. Thus, it is not just disposability that we should be concerned with. We also need to focus on how life is being produced *for* the productive energies and projects of a global economic system that is increasingly relying on biocapitalism as a promissory model of production. Put another way, in order to get the precise type of GE corn Monsanto desires, they need to utilize segments of "junk" DNA material to produce a plant capable of resisting pesticides laced with Agent Orange. Even suboptimal biocapital has a place in the developmental motor of biocapitalism. Schools focused on maximizing the value of human capital (or biocapital) perform a similar process in the era of biocapitalism; it is equally a matter of optimizing the existing stock of human capital by establishing a socio-ecology where monocropping the minds and bodies of students is the natural state of things. Low-yield human capital students and populations as well as high-yield ones are all part of the same artificial ecology that is regulated by the laws of human capital accumulation.

I want to propose here a preliminary framework that may provide exit points from the strengthening union between biocapitalism and education that I have been charting in this book. Following Michael Hardt and Antonio Negri, who have theorized a multitudinal or horizontal model of politics that forms itself in opposition to global neoliberal forms of governmentality, I contend that what is needed is an educational practice and theory of "biopolitical reason" that centers life in a politics of education. By life I mean the social existence that is at stake in human capital models of education where *Homo economicus* is sovereign. I also mean it to signify the productive energies (human vitality and creativity) of individuals and populations that are targeted under biocapitalist educational regimes such as neo-*Sputnik* science education reform policies, or, biochemical governance over the ADHD body in school. Life, as I have understood it in this book, is also the genetic structures and information that have become part of an emerging educational technology of control that uses things such as brain imaging and psychotropic drugs to redirect vital energies in a more efficient and value-enhancing way. Finally, I also use the term life to designate living material outside of the human body and to include the nonhuman, the ways plants, animals, chemicals, soil, rivers, and genetic information, for instance, are a part of the productive and consumptive apparatuses of biocapital of which educational sites are integral and co-constituting. Biocapitalism requires knowledge surplus in the areas of immaterial labor (software programming, imaging rendering, lab technicians, and so on) and as such, educational institutions are deeply implicated in the development of knowledge and practices that contribute to the expanding forces of marketization into the nonhuman world. The reason I see for Hardt and Negri's desire to create a democratic politics where social life is rooted in an alternative biopolitical reason is because it requires that all of these forms of life above and many more are constitutive to the development of a resistant politics that refuses the terms of life biocapitalism has established. Let me offer what Hardt and Negri see as constituting the character of an alternative biopolitical reason so that it will become clearer why I think educational spaces in the age of biocapitalism should be experimenting toward such a model.

For Hardt and Negri, biopolitical reason is the basis by which social life can be produced in an alternative biopolitical manner to those created through Empire or forms of neoliberal governmentality (what produces the human capital educational subject, for instance). Biopolitical reason, as Hardt and Negri articulate the concept, is guided by two principles that they consider to be foundational to the development of resistive practices in social spaces shaped by global neoliberal structures of power. The first principle they assert is the necessity for biopolitical reason to emerge out of "the experience of the common" (Hardt and Negri 2009, 12). What they mean by an experience of the common is "the *production and productivity* of the common through collective social practice. Like the universal, the common lays claim to truth, but rather than descending from above, this truth is constructed from below" (Hardt and Negri 2009, 121). The second connecting principle guiding Hardt and Negri's notion of an alternative biopolitical reason is that knowledge production or political epistemology needs to be "grounded on the terrain of struggle—struggle that not only drives the critique of the present reality of domination but also animates the constitution of another reality" (Ibid.). Here, Hardt and Negri draw on feminist traditions, for example, that emphasize the inherent patriarchal and ethic of domination over the natural world that Western scientific notions of objectivity and truth have promoted. Feminist standpoint theorists—Sandra Harding, Donna Haraway, Patricia Hill Collins, and Nancy Hartsock, in particular—have been making this point for the past 30 years (Harding 2003). According to Hardt and

Negri, just as feminist standpoint theorists have suggested, it is in the act of resistance and critique against the objectivity of science, and the politics of truth it supports in capitalist, patriarchal, and imperial projects where alternative epistemologies can and should be produced. In particular, Hardt and Negri are interested how in such epistemological spaces "a common subject is formed here that has nothing to do with the transcendental" (Hardt and Negri 2009, 121).

The organizing concept of the common is key to Hardt and Negri's understanding of biopolitical reason that I have partly laid out in the preceding text. For them, the common is a produced space of resistance wherein difference (i.e., an individual's race, sexual orientation, or gender) is not absorbed into a unitary structure such as a political party or a universal truth, but rather diversity becomes the motor in which common spaces come into being against global forces seeking to homogenize and enlist productive energies into imperial projects (Hardt and Negri 2004). The common for Hardt and Negri, however, also includes other forms of life that are also targeted as sites of enclosure. Here land, ecological systems, and nonhumans are all considered part of the common and potential zones where individuals and groups can create networked forms of resistance, and thus provisional and emerging constitutions of the common. From the two principles of biopolitical reason I just delineated (experience of the common and an epistemological model from below or an insurgent model of knowledge production), Hardt and Negri offer these provisional three characteristics of biopolitical reason: "Put[ting] rationality at the service of life; technique at the service of ecological needs, where by ecological we mean not simply the preservation of nature but the development and reproduction of 'social relations'... between humans and nonhumans; and the accumulation of wealth at the service of the common" (Ibid., 125).

So how does Hardt and Negri's provisional definition of an alternative biopolitical reason provide a useful jumping-off point for developing educational practices against biocapitalism? I will take each characteristic listed above in turn and point to both its strengths and limitations for developing an alternative biopolitical reason to that of the one operating in biocapitalist modes of education.

As I have argued throughout the chapters of this book, part of what makes biocapitalism unique to industrial capitalism, for example, is the productive need for new (and newly "discovered") forms of life to be enlisted in the circuits of market exchange and consumption. The AquAdvantage® Salmon, the nonhuman I drew upon in chapter 4 to consider an alternative scientific literacy that is open to including bioengineered fish within a more robust democratic politics of education, is a prime example of a form of biocapital that could only come into being through the appropriation of genetic science and two species of fish (the Chinook salmon and the Ocean pout) through biocapitalist production processes. In the example of the first genetically engineered animal made for human consumption, nature (salmon and Ocean pout) is set against culture (genetic science) in the productive regime of biocapitalism. Here reason, in the form of technoscience and the governing rationality that organizes a biocapitalist productive entity such as the AquaBounty Corporation, is in the service of *enclosure* and not the common. Yet, for this very same reason, something like the AquAdvantage® Salmon is a potential pedagogical site where both the foundational principles of an alternative biopolitical reason for Hardt and Negri can be identified: the controversy of the "Franken fish" creates a common experience (in that it unites a variety of political opposition movements such as environmentalists, food sovereignty, antiglobalization, indigenous groups, and so on) and a zone from which an epistemology can be grounded on a terrain of struggle. So in the very heart of biocapitalism, the creation of a new commodity species, an alternative biopolitical reason could emerge that reverses the polarity of power *over* life and turns it into a power *for* life. In this sense, Hardt and Negri's notion of biopolitical reason is actually pushed forward in that nonhumans that were created through a negative biopolitical rationality (the biopolitical reason at work in the AquaBounty laboratories, for example) are transformed into a common nodal point of resistance that brings together a variety of actors into a network of resistance. The Franken fish should not be shunned, in other words, but become a pedagogical nonhuman that can teach humans how to develop and practice an alternative biopolitical reason—one where rationality is "put at the service of life" in that it forces us to question and critique the form of life biocapitalism has set loose in society.

From the example of the AquAdvantage® Salmon, we can also see where Hardt and Negri's second characteristic of an alternative biopolitical reason might emerge: technique or technology at the service of ecological needs. While this is certainly not a new insight that technology and science need to be attuned and connected to a sustainable culture and politics instead of in the service of irrational and insane models of global capitalist development, it is nonetheless an important component to the development of an alternative biopolitical reason.[7] Here again we can focus on how biopolitical reason infused with biocapitalist needs forms regimes of knowledge (genetic science, for example) and practices (reconfiguring the genetic structures of two species of fish) that are against ecological needs for balance and diversity, and instead promote standardization and the regulation of biological material for the productive forces of biocapitalism.

As I argued in chapter 5, we can see how technique and science turned against humans and nonhumans create socio-ecological spaces where life, at least in the example of ADHD students, is under the management of what I have called biochemical governance. Educational life, in this case, is subjected to a technique that is not about creating the best socio-ecological setting (in the school, which is also another example of a negative technology) but rather interested in regulating existing life in a way that best maximizes biocapitalist educational imperatives: optimizing individuals and populations in a way that maximizes the surplus of immaterial labor skills and abilities in society. The primary nonhuman in play in the example of biochemical governance and the ADHD student is the biochemical drug itself (amphetamine in the case of drugs such as Adderall and Ritalin), which is a synthetic chemical that was originally derived from plants (as are most pharmaceutical drugs) to treat ailments such as asthma. With a focus on how technologies and science are utilized in something like biochemical governance, we can see that an alternative biopolitical reason would need to draw attention to and resist how biocapitalism works against the basic principles of ecology (balance, diversity, regenerative systems, etc.) by turning abstracted pieces from different ecologies (plants and students in educational settings) against each other in the mutual service of biocapitalist educational needs. The assemblage of drugs, student bodies in schools, and biochemical governance would be another entry point for developing an alternative biopolitical reason where parents, teachers, and communities could create a common where biochemical governance is refused and rejected as a normalized practice in schools. The technique or technology of biochemical governance itself would be the catalysts for creating epistemologies rooted in common struggle and the production of subjectivities that do not adhere to hierarchical forms of power such as those associated with biochemical governance which I outlined in chapter 5.

Finally, and perhaps the most crucial area where an alternative biopolitical reason needs to focus in the age of educational biocapital is on a rejection of the valorization process created through human capital models of education. As I have argued in chapters 1 and 2, human capital models of education function by individuating educational subjects and measuring their investment and entrepreneurial capacities. Here, education in the biocapitalist era is clearly against "the accumulation of wealth at the service of the common" in that human capital production through education is built around an understanding of personal wealth accumulation, namely, forms of human capital. Moreover, human capital models of education are also constructed around forms of governmentality that compel individuals and groups to make decisions based on the laws of rational economic thinking—*Homo economicus* does not care about his fellow travelers in the world: he is only concerned about himself and how to leverage his own social advantages gained through accumulating the most human capital possible. Here again we can see that the dominant biopolitical reason produced through biocapitalist forms of education are in opposition to the common—in fact, they rely on a pedagogical model that fractures and segments social life through the accumulation process of human capital and the institutions that deliver the investments necessary for living and learning alone with everyone else. At this third level of biopolitical reason, the focus on how wealth is accumulated in society, we can see that education must be freed from its participation in "organiz[ing] the environment as a cage for *Homo economicus*" (Illich 1981, 12).

The question remains however: How can educational spaces (both formal and nonformal) participate and, in fact, produce alternative biopolitical rationalities? This is a particularly weighted question since, as I have pointed out in the chapters of this book, educational sites are absolutely crucial to the development and progression of biocapitalism as an economic and political model. I will lay out in the following section what I believe are necessary components to educational practices that can potentially serve as points of departure for developing alternative biopolitical rationalities:

1. Curriculum and research based on what Hardt and Negri call *strategic investigation* or collective practice;
2. the establishment of pedagogical relations with the nonhuman world;
3. a politics of education built on a redefined notion of civics.

Let me take each of these provisional principles in turn as each are distinct but also interrelated. The first principle I suggest is not necessarily a new proposal. Participatory action research, for example, is a method by which knowledge production is rooted in collective investigation around a community or public problem—one where researcher(s) equally contribute to a viable solution to an agreed upon problem. What I am suggesting is that instead of being a marginalized mode of knowledge production, something like participatory action research that privileges knowledge from below needs to become the foundation of educational experience. The key here would be that instead of the curriculum being a technology of control, a way to manage how and what knowledge is produced in educational spaces, a curriculum based in strategic investigation would identify community problems from a model of "co-research" that could "construct together with workers [or students and community members] alternative knowledges from below that are completely internal to the situation and intervene in the current power relations" (Hardt and Negri 2009, 127). Again, the suggestion to root knowledge production in such a strategic investigation model is not a utopian or idealistic dream; there are existing models to draw upon and experiment with such as the International Science Shop Network in Europe where university scientists work alongside community members and contribute and share research to solve problems such as water pollution.[3] Existing collective research models are also in place in many local community garden organizations that have created educational communities rooted in sustainable and food justice concerns. There are a variety of nonformal educational spaces that can be looked to as resources for developing strategic investigation models of learning and creating knowledge from below.

The next related provisional principle that I am suggesting is important to the development of an alternative biopolitical rationality of education involves the construction of educational approaches that can create a pedagogical relationship between the human and nonhuman worlds. Part of the problem with current biocapitalist models of education is that they perpetuate the myth that nonhumans belong to the natural world and, therefore, should be treated and understood as something to be dealt with through the mediator of science. In other words, if scientific experts working on behalf of biocapitalist interests are the only ones (or at least the most influential at the juridical level) who can tell us how we should understand and learn about nonhumans, then we also are accepting the power relation that scientific and technological discourses and practices are the arbiters for determining whether or not nonhumans are part of democratic politics. Here again we need to focus on the ways technique is being used in the service of biocapital and not for life in common. One educational practice that I think is particularly useful for reversing the biopolitical reason promoted through biocapitalism is the creation of actor network maps (as I discuss in chapter 4). The reason actor network maps can be thought of as a technology or technique in the service of ecological needs is because they can help us better understand the social relations between humans and nonhumans, and *how* they are constructed. Thus, in order to get toward a common experience with nonhumans, we need educational spaces and practices that can chart and understand the ways things such as the ADHD epidemic and genetically engineered animals are connected to shifting networks of power that must be understood and brought into democratic conceptualizations of education.

Finally, the modern notion of civic education needs to be reconstructed in order for an alternative biopolitical reason to emerge within educational spaces. Traditional notions of democracy that have been used as educational staples are built around ideas and concepts that continue to separate nature and culture and, therefore, are in alignment with biocapitalist configurations of oppressive biopolitical rationalities. For example, the effects of environmental racism on a community are usually rooted in an understanding of civics that turns to governmental or state intervention for alleviating the accumulation of toxic dangers that historically have been disproportionately dumped on the working class and communities of color (Bullard 2000). In such hierarchical power relations, entities such as highly polluting corporations have essentially unlimited power in how the problem of environmental racism is managed. Democracy is mediated through the established power relations vested in corporations and the biopolitical rationality they tend to exact on forms of life: one only concerned with value maximization and how to most cost effectively distribute externalities. Yet perhaps most importantly what is also maintained in this model of corporate democracy is the view that environmental toxins are nonhumans that can only be understood through the gatekeeper of an expert—effectively creating the illusion of a one-way street between the nonhuman (arsenic, for example) and its only legitimate communicator: the scientist. What is at stake here is what and who we consider to be actors that shape our socio-ecologial realities. In other words, from who and what should we be learning about nonhumans as important as arsenic that intervene in the health of individuals and communities? I am suggesting that a traditional framework of civic education roots democracy in a nature/culture binary that makes it impossible to create truly democratic relations with nonhumans that constitute important relations of our social realities. One of the places where we can see a model of democratic politics that centers nonhumans is in the food justice and food sovereignty movements. Here, nonhumans such as GMO seeds are the pivot point for a resistive mode of democratic politics that connects genetically engineered seeds to larger networks of power that threaten sustainable and more just relations with people, land, and plants. In this sense, developing a reconstructed concept of civics in educational settings that rejects the culture/nature binary will most likely depend on the ability to expand what is considered education so that learning relationships can be established with movements and community groups that are already equally considering nonhumans in their model of democratic politics.

Of course, these principles are only starting points for developing alternative biopolitical rationalities in educational settings (which I consider spaces like local slow food movements or learning garden networks also to be). I do, however, believe that they are good anchoring points to build out from. What is ultimately at stake, however, is the future of education, and whether or not it will be in the service of life or in the service of extracting value from life. Naming the ways in which biocapitalism is determining the terms by which we understand, learn about, and are part of such a productive and consumptive process is only the first step. The next is to refuse these conditions and create alternative ones by experimenting with exit points within educational spaces that the age of biocapitalism has now forced us to consider.

Notes

Introduction

1. It is interesting to note the connection between Foucault's point that the Nazi state was in fact not idiosyncratic to the modern period, but rather an expression of sovereign power's full logic and Max Horkheimer and Theodor Adorno's (1947) earlier thesis they made in their magisterial *Dialectic of Enlightenment*. That is, the triumph of instrumental reason, for Horkheimer and Adorno, was a result of the cultural logic set in motion by the advanced capitalist state. For them, Auschwitz and the Nazi state also stood as the grim apex of the state's ability to totally administer all dimensions of life.
2. As Thomas Lemke (2011) has noted, Foucault's usage of the terms biopower and biopolitics is not neatly distinguished. However, at the most general level, biopower in Foucault's thought emerges through its differentiation from sovereign forms of power that existed in the early modern period that mostly rested on juridical grounding. In other words, "Foucault sees the particularity of this biopower in the fact that it fosters life or disallows it to the point of death, whereas sovereign power takes life or lets live. Repressive power over death is subordinated to a power over life that deals with living beings rather than with legal subjects" (36). Biopolitics therefore is related to biopower, in that once biopower reached a level of control within a state's population (from the eighteenth century forward for Foucault), it achieves the status of biopolitical—or a politics of society based on the administration, control, and regulation of the life of a population.
3. It is important to note that Du Bois was one of the earliest political theorists to name the power relations associated with white supremacy. His most sustained discussion of the term in different contexts is probably in his collection of essays titled *Dark Water: Voices from within the Veil* (2011). White supremacy is a significant early biopolitical concept in the work of Du Bois as it recognizes and articulates how inferior and superior subsets of human subjectivities are produced within US society and culture, as well as the ways in which these differences are governed. Since Du Bois, Charles Mills has perhaps best described the intellectual (epistemological) and the material components that constitute white supremacy and originate in what Mills has identified as "The Racial Contract." The Racial Contract "is that set of formal or informal agreements or meta-agreements (higher-level contracts *about* contracts, which set the limits of the contracts' validity) between the members of one subset of humans, henceforth designated by shifting 'racial' (phenotypical/genealogical/cultural) criteria C1, c2, c3, . . . as 'white,' and

coextensive (making due allowance for gender differentiation) with the class of full persons, to categorize the remaining subset of humans as 'nonwhite' and of a different and inferior moral status, subpersons, so that they have a subordinate civil standing in the white or white-ruled polities the whites either already inhabit or establish or in transactions as aliens with these polities, and the moral and juridical rules normally regulating the behavior or whites in their dealings with one another either do not apply at all with dealings with nonwhites or apply only in a qualified form (depending in part on changing historical circumstances and what particular variety of nonwhite is involved), but in any case the general purpose of the Contract is always the differential privileging of whites as a group with respect to the nonwhites as a group, the exploitation of their bodies, land, and resources, and the denial of equal socioeconomic opportunities to them"(Mills 1997, 11). I see Du Bois recognizing how such an understanding of white supremacy is constitutive to caste education and thus its biopolitical underpinnings.

4. C. S. Coon's influential study on racial classification in his *The Origin of Races*, for example, presented a scientific argument for evolutionary subsets within the human population along phenotypic lines as late as 1962. For Coon, the five racially distinct groups of humans Australoid, Mongoloid, Caucasoid, Congoid, and Capoid were supported by eugenic science methods such as craniometry and other types of body measurement. Coon's racial science was quite mainstream and in fact perpetuated eugenic racial science claims for superior and inferior types of humans, a major area of political struggle that Du Bois vehemently opposed throughout his career. Du Bois saw eugenic science and the politics of race that grew out of it as a form of rationality that supported and maintained caste relations in the United States. For another brilliant critique of eugenic science and its connection to the political ordering of the state, see Georg Lukács' excellent analysis of the racial science underpinning the Nazi state in his *Destruction of Reason*.
5. The Reconstruction period following the Civil War was a particularly important period for Du Bois. It is during this period that Du Bois (1999) argued that the United States missed an opportunity to truly create a democratic society through what he called "abolition democracy," or the abolition of insitutions and political structures that maintained white supremacy's hold on citizenship. What Du Bois lamented most about the Reconstruction period in US history was the missed opportunity to create social institutions and legal apparatuses that abolished citizenship based on a white supremacist basis. Also see Angela Davis's (2005) excellent discussion of Du Bois's notion of abolition democracy and Joel Olson's (2004) analysis of this aspect of Du Bois's work.
6. See Ulrich Bröckling's (2011) excellent account of the Foucault's analysis of the Ruldolf Goldscheid's early economic theory of human capital that preceded the later American version of the Chicago School.
7. As Gordon (1991) has pointed out, Foucault's understanding of the state is a decidedly non-modern one. That is, the state for Foucault can better be understood as a "function of changes in practices of government, rather than the converse" (4). In others words, it is the practices and techniques of government that best explain the entity that modern political theory has traditionally termed the state. Or, it is the practices and flexible strategies of ruling over the conduct of individuals in a given population through the state that is important, not necessarily the Leviathan itself.
8. In her excellent study on forms of animal capital, Nicole Shukin (2009) has accurately noted that Hardt and Negri's theory of biopolitical production under Empire and the variety of governmental strategies it employs leave out entirely an analysis of

nonhumans. This is an important insight into one serious limitation of Hardt and Negri's analysis, that is, the ways in which nature, animals, and plants, for example, have been wrapped up in modes of production and consumption ever since the colonial age, is not given consideration. In agreement with Shukin, I problematize this oversight in Hardt and Negri's biopolitics by arguing that life in general needs to be the loci of biopolitical analysis in our current postgenomic moment.

9. According to the United Nations Food and Agriculture Organization' (2008) report *The State of World Fisheries and Aquaculture 2008*, "world aquaculture has grown dramatically in the last 50 years. From a production of less than 1 million tonnes in the early 1950's, production in 2006 was reported to have risen to 51.7 million tonnes, with a value of US $78.8 billion. This means that aquaculture continues to grow more rapidly than other animal food-producing sectors. While capture fisheries production stopped growing in around mid-1980, the aquaculture sector has maintained an average annual growth rate of 8.7 percent worldwide since 1970" (6).
10. Not surprisingly, over the past 10 years, *New York Times* editorialist and author Thomas Friedman has become one of the most revered educational policy analyst in the country. His flat world thesis, for example, makes up the theoretical basis for the National Academies of Sciences (2008) report *Rising Above the Gathering Storm* that Friedman (2007) himself lauds as being "put together by a blue-ribbon panel of scientists and entrepreneurs, concluded that for America to be prepared for the twenty-first century, it must increase federal investment in such research [science, mathematics, and engineering] *by 10 percent a year over the next seven years*" (362). Friedman thus sees the economic battle in the flat world as taking place between countries such as China and the United States where China is "now focusing on how to unleash more creative, innovative juices among its youth,"while kids in the United States are quickly lagging behind.
11. Bolivia's recent forms of resistance to the global hegemony of Western developmental models based on exploitation of the earth and the poor may perhaps be the best example of what Hardt and Negri might consider a form of governmentality that demonstrates strong tendencies of "biopolitical reason." From their recent alternative global summit on climate change that was far more progressive than Copenhagen's market-based approach, Bolivia, along with Ecuador which has already passed such legislation in its constitution, is currently proposing a UN treaty to include the rights of nature as part of its humanitarian (extending it to the nonhuman) mandate. In many ways, Bolivia and Ecuador's move to include ecosystems and land as part of the political constitutions of their country represents a reclaiming of the commons, a central goal to Hardt and Negri's multitudinal politics.
12. Donna Haraway (2008), one of the first theorists to begin charting the terrain of biocapital, has also posed its development in this way in her studies on human/nonhuman species relations: "If a Marx equivalent were writing *Biocaptial*, volume I today, insofar as dogs in the United States are commodities as well as consumers, the analyst would have to examine a tripartite structure: use value, exchange value, and encounter value, without the problematic solace of human exceptionalism" (46).
13. In this area of biocapitalist studies, physicist and activist Vandana Shiva and her groundbreaking work on biopiracy must also be included as a foundational text (Shiva 1997).
14. Drawing on Karl Marx's observations of the way credit and debiting function as a means to extend the valorization process beyond the material world and into the fantastical in the *Grundrisse*, Cooper also sees biocapitalism as being built upon a

similar fictitious ground. Biocapitalism for Cooper (2008) is thus part of a larger economic and political project where "fueling this apparently precarious situation is the delirium of the debt form, which in effect enables capital to reproduce itself in a realm of pure promise, in excess of the earth's actual limits, at least for a while. This is a delirium that operates between the poles of utter exhaustion and manic overproduction, premature obsolescence and the promise of surplus" (31).

15. As Helmreich points out, Marxist feminists have argued that Fredrich Engels' (2010) *The Origin of the Family, Private Property and the State* has had an important influence on thinking about the ways capitalism is also invested in managing the reproductive capacities of the working class. Though still mired in a patriarchal view of family, largely the result of using Lewis Henry Morgan's work in tracing familial relations in different historical periods, Engels' work on the family is important in that it recognized human reproduction as integral to capitalist development. Feminist kinship studies, through their critique and challenge of traditional kinship understandings in the field of anthropology that essentialized categories such as nature, gender, and reproductive pairings within the pre-determined biological framework of modernity, have been at the forefront of theorizing how technoscientific advances have fundamentally changed debates around gender, reproduction, or conceptions of familial relations. For an excellent collection of work in the area of feminist kinship studies, see Franklin and McKinnon (2001).
16. Reprogenetics is the term created by Princeton molecular biologist Lee M. Silver (1997) who envisions in the not-to-distant future the merging of reproductive and genetic technologies in ways that will be able to manipulate genetic material from numerous parents for desired traits in their offspring.
17. There are also a number of excellent works in the Foucault literature in education that have helped pave the way for a broader theory of educational biopolitics. See, for example, Ball (1990), Marshall (1995), Popkewitz and Brennan (1997), and Olssen (1999).
18. Also see Lewis's (2007) where he approaches the utopian debate in education along biopolitical lines. Building off Foucault and Hardt and Negri, Lewis seeks here to rethink the notion of multitudinal democracy through the figure of John Dewey.
19. Within the literature on biopolitics and education, Lewis and Kahn's (2010) wonderfully bazaar book *Education Out of Bounds: Reimagining Cultural Studies for a Posthuman Age* must also be seen as having a deeply biopolitical project in their argument for reimagining educational politics and possibilities through the figure of the monster and exopolitical communities.
20. Also see Ball and Junemann (2012) for an excellent ethnographic account of neoliberal governmentalities of education in Europe.
21. See Pauline Lipman's (2002) excellent study of school reform in Chicago driven by "corporate, financial and political elites" that according to her analysis are creating the conditions where "current policies actually exacerbate existing inequalities and create new dynamics of inequality with important implications for students and for the future of the city as a whole" (380).
22. For an analysis of the connection between Marcuse's notion of "productive destruction" and biopolitics, see Kellner, Lewis, and Pierce's introduction to *Herbert Marcuse: Philosophy, Psychoanalysis, and Emancipation* (2011).
23. Perhaps the most egregious example to date of a neoliberal utopian dream in the higher education is Ashford University in Clinton, Iowa. Designed by Bridgepoint Education Inc., a publically traded commodity started by former executives of the University of Phoenix, the entire institution functions almost completely as a

mechanism for capturing federally funded student loan money. After purchasing the tiny 300-student school from a group of Franciscan nuns for its national college accreditation, the Bridgepoint Education Inc. now "tap[s] federal financial aid dollars, the source of nearly 85 percent of the university's revenues—more than $600 million in the last academic year. Ashford now counts nearly 76,000 students, 99 percent of whom take classes online" (Kirkham 2011).

24. In a recent interview, Jacob Lew, Obama's Office of Management and Budget Director, laid out the new regulatory framework that will allow student loan interest to begin accruing while students are still enrolled in graduate programs. The logic behind the financial giveaway to corporate finance is part of an overall plan to save money in the federal budget by reducing federal subsidies for graduate student loans. See the transcripts of Lew's comments on the new student loan interest plan at http://transcripts.cnn.com/TRANSCRIPTS/1102/13/sotu.01.html
25. In Peters et al. (2009), a very impressive range of studies on forms of educational governmentality is offered. While two essays in this edited volume focus on themes such as knowledge capitalism and higher education as a neoliberal form of governmentality aimed at accruing "capital for national competitiveness in a global economy" and another on the effect of brain-based learning as a form of understanding the educational subject (Wong 2009), there is no discussion on ways in which biocapitalism intersects with educational governmentality.
26. Postgenomic, according to Sarah Franklin and Margaret Lock (2003), is a term derived from "the hype surrounding the project to map the human genome [which] has already give[n] way to a new phase of enthusiasm for what has become known as postgenomics. It includes the new science of proteomics, in which an overriding interest in the gene has largely been superseded by interest in complex models of protein and cellular interactions. In connection with developments such as cloning, and in the emergent science of tissue engineering, the cell has reemerged as a central unit of action, temporally and spatially as well as functionally" (13).

1 Learning to be *Homo economicus* on the Plantation: A Brief History of Human Capital Metrics

1. Recently, teachers in the state of Hawaii voted to reject teacher performance measures that the state was forced to adopt as recipients of Race to the Top funding. This is the first time to date that a state has refused the mandates required by Race to the Top legislation and it remains to be seen what the Department of Education will do in response (See Strauss 2012).
2. Foucault's explanation for the United States's more radical configuration of neoliberal economic theory and governmental approaches compared to earlier German and French versions stems from the fact that liberalism served as the philosophical and political founding of the United States as a modern nation-state. That is, different from Western European nations such as Germany, France, or England, the United States's first political framework was deeply rooted in liberal notions of free-market and commerce ideologies of democracy. It is for this reason that Foucault saw very little differentiation between the major political parties in the United States when it came to the development of neoliberal forms of governmentality (Foucault 2008).
3. Claudia Goldin and Lawrence Katz's book *The Race between Education and Technology* has a direct connection to the Chicago School's neoliberal tradition. Goldin was mentored and studied under both Robert Fogel and Gary Becker during her graduate training at the University of Chicago's economic department.

Her recent book along with coauthor Katz applies a similar human capital analysis to the relationship between technological advancement and education over the twentieth century. For Goldin and Katz, there has been a sharp decline since the late 1960s in the productive output of educational institutions in terms of how it feeds technological breakthroughs in key areas of the economy. The question of equity in the book is also treated similar to the way Fogel and Engerman do in *TC*. In other words, equity is a question of the human capital accumulation ability of the individual as well as the overall population, and not one of historic and political structures that have defined in the United States how different people and groups access and experience educational institutions.

4. In Foucault's examination of neoliberal economic theory, the figure of *Homo oeconomicus* serves a very important function. Foucault's charting of the domain of *Homo oeconomicus* begins with Adam Smith and utilitarian thinkers Jeremy Bentham and John Stuart Mills' shared view that all rational human beings were, at their core, economic beings acting in their own self-interest. Foucault sees neoliberal theorists, especially those of the Chicago School, as expanding this original understanding of *Homo oeconomicus*, the rationality of economic conduct, to include all domains of life. Foucualt states, "the most radical of the American neoliberals, if you like—says that it is still not sufficient, that the object of economic analysis can be extended even beyond rational conduct as defined and understood in the way I have just described [neo-classical conceptions], and that economic laws and economic analysis can perfectly well be applied to non-rational conduct, that is to say, to conduct which does not seek at all, or, at any rate, not only to optimize the allocation of scarce resources to a determinate end." This expanded understanding of *Homo oeconomicus* situates even the most "irrational" or personal decisions to economic rationality such as marriage, child rearing, education, health decisions, and so on; however, it also operates in a larger social field of action that Foucault (2008) argues makes *Homo oeconomicus* an even more "governmentizable subject": "*Homo oeconomicus* is someone who accepts reality. Rational conduct is any conduct which is sensitive to modifications in the variables of the environment and which responds to this in a non-random way, in a systematic way, and economics can therefore be defined as the science of the systematic nature of responses to environmental variables" (269). Also see Bröckling (2011) for a very incisive account of the figure of *Homo oeconomicus* in Foucault's critique of human capital.
5. Harvey Kantor and Robert Lowe have pointed out recently that human capital arguments in educational reform debates stretch back all the way to Horace Mann's 1841 appeal to the Massachusetts Board of Education that highlighted the economic value of a common school system (Kantor and Lowe 2011).
6. Robert Fogel has never abandoned his cliometric research on slave plantations, in particular, how slave plantations in the antebellum south have wrongly been interpreted by previous economic historians as nonproductive economic zones. See his (1989) *Without Consent or Contract: The Rise and Fall of American Slavery* where he amends some of his and Engerman's earlier data points from *Time on the Cross* but ultimately argues that slave life was indeed productive and educative in a market sense.
7. One of the keenest early critics of human capital theory who often gets overlooked was Ivan Illich. In his wonderful book *Shadow Work*, Illich investigates the usurping of the commons (he uses the term vernacular which designates subsistence cultures and nonmarket social relations) by premodern and modern institutions such as

the Catholic Church. Ultimately, Illich argues that subsistence cultures underwent a massive and systematic standardized transformation through industrialization. For him, industrial society is characterized by its tendency "to organize the environment as a cage for *homo economicus*" (Illich 1981, 12). For Illich education, namely, formal schooling, played a central role in the process, and instead of being places where learning occurred for all, they are places where "students ask if they are in school to learn or to collaborate in their own stupefaction" (Illich 1981, 31).

8. Fogel and Engerman's study however, should not be seen as an outlier in the field of economics. Rather, it was indicative of a rising tide of a "new economic history" based on "cliometrics" that utterly transformed the field of economics starting in the 1970's. Their *Time on the Cross*, for example, originally took its inspiration from a paper cowritten by Alfred Conrad and John Meyer (1958) which was "a major spark to the cliometric revolution" (Goldin 1995, 193). The cliometric revolution was seen at its inception as a major methodological research breakthrough in the field of economics. See Claudia Goldin's (1995) essay which gives high praise to Robert Fogel's role in the development of cliometrics that also helped him achieve a Nobel Prize in Economics.
9. The similarity between human capital theorists' "scientific revolution" that allowed a reinterpretation of economic history through quantitative methods such as cliometrics and pseudo-scientific methods of the eugenic era such as craniometry is striking. Both models claim to offer objective and nonbiased truth claims about racial difference based on human biology in the case of craniometry and on the actual condition of work on the slave plantation of the antebellum south through cliometric analysis. In both cases, the socially constructed nature of race is neutralized through models of quantitative analysis and what is promoted instead is a biopolitical project of maintaining racial difference as a form of biopower: one as eugenic science and the other through continued economic and social forms of inequality that are erased through human capital narratives.
10. There is another productive way to think about Fogel and Engerman's human capital argument worth noting. Thought of in a classic Marxist dialectic framework, what Fogel and Engerman are attempting to achieve in their analysis of slave labor and life is a short-circuiting of the slave–master dialectic. That is, what Fogel and Engerman want their readers to believe is that the moment of true learning from a slave–master relationship does not come from a recognition of an unequal power relation and the desire to destroy it, but rather an embrace of the master's ideology, values, and habits. In other words, historical progress does not emerge through a critical resistive moment from the slave, but rather by the full assimilation of the society and culture of the master.
11. The concept of human beings as economic investment machines, the basis of human capital theory, cannot be viewed outside of its analog theory of human health. That is, investments in the human capital framework can only be measured and qualified against a definition of an individual's income stream—the return people receive for their work which correlates directly with the amount of human capital one has accumulated over a life cycle. In this sense, human capital theory, as Foucault recognized 30 years ago, is nothing if not biopolitical. Human life, how long one lives and the quality of life one experiences (in terms of access to health care, education, high-paying employment, safe and environmentally clean housing, etc.), is truly a matter of how well one can master skills and strategies for attaining a high level of human capital accumulation. Bröckling makes a similar point in his analysis of economic theories based on human capital models.

12. For one of the best studies of slave rebellions and their political meaning that takes Du Bois as a point of departure, see Paul Gilroy's (1993) *The Black Atlantic: Modernity and Double-Consciousness.*

2 Schooling for Value-Added Life: The Making of Educational Biocapital

1. Since the publication of the *Los Angeles Times'* value-added story, Derek Briggs and Ben Domingue (2011) have provided a critical analysis demonstrating how the statistical models used by the RAND economist Richard Buddin hired by the *Los Angeles Times* to create the performance scores were deeply flawed.
2. The adoption of value-added models in the Department of Education's push to integrate value-added assessment models into state reform measures has almost completely ignored studies from major research centers that show the flaws of using such models as a tool to guide large-scale educational reform. See, for example, the RAND Corporation's study Evaluating Value-Added Models for Teacher Accountability (McCaffery et al. 2005) or the National Research Council's Board and Testing and Assessment Letter Report to the US Department of Education (2009). Also see Linda Darling-Hammond's (2012) report on creating a comprehensive system for evaluating and supporting effective teaching that is largely a response (albeit a mild one) to value-added techniques.
3. One of the more famous cases in the United States where value-added measures were used to "clean house" was the Washington DC school district headed by the lightning rod Chancellor Michelle Rhee. Rhee, using value-added scores, fired 241 teachers (and counting) in the district and has been one of its most vocal proponents, appearing in films such as *Waiting for Superman* where she advocates for value-added measures as a panacea for real educational reform in the United States (Turque 2010).
4. Jason Kamras, the director of the "human capital strategy for teachers" for the Washington DC District, was an important figure in the DC value-added reform push. He played an integral part in getting the DC teacher union to accept the new value-added-driven contract and was pointed to by George W. Bush as national teacher of the year in 2005, and is considered by many to be part of the new face of educational reform in the United States.
5. In early 2012, the *New York Times* published the rankings of the New York City School District's teacher value-added scores, causing a massive media spectacle in the form of a public shaming ritual. It is quite clear that one of the strategies of value-added proponents is to create media spectacles that depict rotten teachers and an ineffectual accountability system as the problem to widespread school success. Even Bill Gates, who along with his partner have invested millions in their own teacher measurement system, has publicly condemned the practice of posting teacher value-added scores in the media (Gates 2012).
6. As Kenneth Saltman has pointed out in his excellent study of the rise and fall of the Edison School, Lamar Alexander was also involved with fellow Tennessean H. Christopher Whittle in business dealings involving privatization schemes in public education such as Channel One and the Edison School (Saltman 2006).
7. After retiring from the directorship of University of Tennessee's Value-added Research and Assessment Center, Sanders is now a research fellow in the University of North Carolina system and is senior manager of value-added assessment and research for SAS

Institute Inc. SAS, according to their website, is "the leader in business analytics software" and provides "services [to] help customers make fact-based decisions to improve performance, from identifying the right product to market forecasting trends."

8. It is interesting to note that some of Stern Stewart & Company's proudest applications of EVA have come in the financial and extractive industries. In discussing the evolution of their EVA metric for firms, the company points out that "among the more sophisticated and value oriented financial institutions EVA® became an attractive approach to measuring and rewarding performance. In mining firms, we discovered that the greatest value in the firm was the reserves in the ground that only trickled slowly through the profit and loss statement over a long period of time. In short, the accounting framework fell far short of the needs of management and the board of directors. Because of the trickle process, their focus was necessarily on short-term results, which is certainly not in the interest of shareholders who are interested in firms maximizing long-term value of the firm. Our firm developed the approach to be used for natural resource firms. Today, extractive industries have no difficulty in using EVA®" (Stern Stewart & Company 2011).
9. Douglas N. Harris, also a member of the Value-added Research Center at the University of Wisconsin-Madison, has recently written the most carefully crafted public relations style-book length treatment of value-added measures in education. Here, Harris attempts to demystify and cut through the cantankerous value-added reform debate by arguing that responsible use of value-added metrics offers states and the federal government the most useful tool for assessing the learning of students and the performance of teachers available. See his *Value-Added Measures in Education: What Every Educator Needs to Know* (2011).
10. The National Center for Performance Incentives (NCPI) housed at Vanderbilt University is also one of the nation's top value-added think tanks. The NCPI states its purpose as "address[ing] one of the most contested questions in public education: Do financial incentives for teachers, administrators, and schools affect the quality of teaching and learning? NCPI's work involves a series of rigorous research initiatives, including randomized field trials and evaluations of existing pay-for-performance programs. We are engaged in these research and development activities to inform both education policy and practice, and to improve teaching and learning within our nation's public schools" (NCPI 2011).
11. Another area where racial resegregation is rapidly increasing through neoliberal restructuring efforts in education is in the charter school system. Charter schools, in cities like New York, for example, have increased racial and class segregation to even greater degrees than the public school system had achieved in these cities. See Mead and Green III 2012.
12. See Pauline Lipman's (2002) excellent work on the neoliberal restructuring of schools in Chicago where she analyzes how economic policies of the city reinscribe historic forms of racial and class inequalities, especially as it relates to educational opportunities for working class and low-income communities of color.
13. In fact, one could argue that there has always been a deep desire in the West to understand and organize politics around heredity. Plato's *Republic*, for example, can be read as a premodern, eugenic-like organizational structure of education and state where one's class and bloodline largely determined one's educational ability and function in the polis. See, for example, Forti's (2006) excellent work on tracing the racial theories of national socialism back to antiquity and, in particular, Plato's articulation of a "Metaphysics of Form."

14. *Harvard Educational Review* (HER) also published (though only a quarter of the length of Jensen's article) a rebuttal to Jensen's article on IQ and scholastic achievement. See Deutsch (1969). It is interesting to note that the late 1960s, the period in which HER published multiple articles in a volume on IQ and school achievement, was also one of the most tumultuous times in US educational reform history when there was a concerted effort to dismantle segregated schooling. In one way, Jensen and other researchers like himself were providing the scientific basis for justifying segregated schooling, in that low IQ students (students of color) were not genetically hardwired for more challenging forms of education and thus segregated schooling was just a natural evolutionary outcome of sound population genetics informed policy.
15. The Social, Genetic, & Developmental Psychiatry Centre housed in King's College of London is considered a world leader in research on behavioral genetics. The center's homepage, for example, boasts that "In the 2008 RAE (Research Assessment Exercise) the sub panel on Psychiatry, Neuroscience and Clinical Psychology made special mention of the Centre twice in its commentary on the Institute of Psychiatry: '. . . evidence of the world-leading outputs especially with respect to the social, genetic and developmental psychiatry (SGDP) . . .' and '. . . the environment was internationally excellent and in the case(s) of social, genetic and developmental psychiatry (SGDP) . . . , world leading" (King's College of London's Institute of Psychiatry, Social, Genetic & Developmental Psychiatry Centre 2012).

3 Engineering Promissory Future(s): Rethinking Scientific Literacy in the Era of Biocapitalism

1. See the Biotechnology Institute's web page where they offer resources for teachers and students as well as information about their BioGENEius competition. Not surprisingly, the board members of the Biotechnology Institute almost entirely consists of CEOs from major pharmaceutical companies. Onilne at: http://www.biotechinstitute.org/.
2. For an excellent study of how the *Sputnik* era economic and national security concerns impacted science education reform and, in particular, how science was taught in the classroom, see Rudolph (2002).
3. I am using the term technoscience to designate what Bruno Latour (1987) and Donna Haraway among other science studies theorists have argued is a conflation of science and technology into a fluid relationship of knowledge production. That is, contemporary science cannot be understood apart from the powerful technological tools that help bring new scientific "discoveries" into existence. Put simply, without a computer and complex software programs, the mapping of the human genome could not have occurred. Thus, the conflation of technology with the activity of science suggests a synergetic relationship as opposed to traditional views of science (knowledge) and technology (application of knowledge).
4. With this said, I am also being careful not to solely rely on Marx's notion of labor. That is, I am adopting Foucault's view that the human capital theory of labor developed by economic theorists at the University of Chicago in the late 1950s and 1960s is one that attempts to sidestep the classical Marxian notion of labor: labor as a form of social power that results from the temporal calculations of capitalist modes of production used for extracting surplus value from human work. It is on this point that I believe Foucault's critique of human capital theory of labor that emphasizes how the processes of extending economic analysis into domains

considered to be noneconomic such as human and nonhuman life go beyond Marx in very fruitful ways (Foucault 2008).

5. Alexander Sidorkin has pointed out elsewhere that the human capital theory of education is based on the erroneous assumption that more education equals higher lifetime income (Sidorkin, 2007). As a strategy to shift away from human capital education, Sidorkin focuses on student labor to disrupt the widely accepted notion that increased investment in education is the only answer. My analysis takes a very different line that focuses on the biopolitical implications of human capital education, that is, the form of and uses of life promoted within a biocapitalist educational context.
6. As Donna Haraway (1997) points out, Sarah Franklin's work on the emergence of genetic material as commodified streams of information was the first to identify how "nature becomes biology, biology becomes genetics, and the whole is instrumentalized in particular forms" (134).
7. Foucault's analysis of neoliberalism in his 1978–1979 College of France lectures also points out the active role that states must take in creating the conditions for free-market economics: 'There will not be the market game, which must be left free, and the domain in which the state begins to intervene, since the market, or rather pure competition, which is the essence of the market, can only appear if it is produced, and produced by an active governmentality" (Foucault 2008, 121).
8. For example, in 2006 the Utah state legislature passed a funding initiative to form the "revolutionary" state/academic partnership the Utah Science and Technology Research (USTAR) initiative. Its goal is to "invest in world-class innovation teams and research facilities at the University of Utah and Utah State University, to create novel technologies that are subsequently commercialized through new business ventures" that "will produce new technologies in multi-billion dollar markets' and where [i]nnovation focus areas include biomedical technology, brain medicine, energy, digital media, imaging technology, and nanotechnology." Online at: http://www.innovationutah.com/aboutustar.html.
9. Nikolas Rose attributes the term "biovalue" to Catherine Waldby who first coined the term in her research on how body parts and tissues of the dead are utilized in ways to enhance the vitality of the living (Waldby 2000). For Rose, "biovalue" can also be used in more general terms to describe "the value extracted from the vital properties of living processes" (Rose 2007, 32). Vandana Shiva must also be seen as an early theorist of "biovalue" in her groundbreaking work on bioprospecting and biopiracy. See, for example, Shiva (1997, 2005).
10. Friedman is by no means the sole theorist to make the claim that human capital education is a key site of crisis to focus on. Following in the footsteps of early US theorists of human capital Gary Becker and Jacob Mincer, Claudia Goldin and Lawrence Katz, two prominent Harvard economists, also rely heavily on a human capital argument in their highly influential book *The Race between Education and Technology*. For Goldin and Katz (2008), the United States' current slippage from global educational and economic leader can largely be explained through this formula: "In the first half of the century, education raced ahead of technology, but later in the century, technology raced ahead of educational gains. The skill bias of technology did not change much across the century, nor did its rate of change. Rather, the sharp rise in inequality was largely due to an educational slowdown" (8). In their framework, investment in education and the development of greater amounts of human capital for a hungry biocapitalism is the solution presented for solving complex educational problems such as social inequality.

11. Friedman focuses on three "dirty little secrets" of the US educational system that, in his estimate, are key sites where transformation must occur. The three secrets he identifies are (1) slipping production rates in science and engineering PhDs compared to that in other countries such as China and India; (2) the decline of worker ambition (one Indian worker, according to Friedman, can do the work of three US workers) in the United States; and (3) the growing overall education gap in the United States that leads to low numbers of individuals who have the requisite skills and knowledge for the relatively sophisticated forms of labor the flattened world demands (Friedman 2005).
12. See Jennifer Baichwal's (2007) excellent documentary film *Manufacturing Landscapes* in which photographer Edward Bertynsky depicts the totality of neoliberalized forms of labor across the globe. Bertynsky's film is a strong counter-example to Friedman's celebratory framing of other nation's work habits and environments.
13. Biomaterials from the "natural" world are not the only ones sought after in biocapitalist markets. As Amit Prasad (2009) brilliantly argues in his research on drug trials in India, test subjects for pharmaceutical drug companies' trials are a highly desired commodity in the biomedical field.
14. The GrowHaus has been influenced by food justice projects such as Will Allen's Milwaukee-based urban farms that focus on the production and distribution of healthy sustainable growing practices in marginalized communities. Online at: http://www.growingpower.org/ for a history of Allen's Growing Power organization.
15. See Richard Kahn's (2010) excellent work on multitudinal democracy and "new science" in his discussion and analysis of the Shundahai Network's peace camp and traditional ecological knowledge.

4 Learning about AquAdvantage® Salmon from an ANT: Actor Network Theory and Education in the Postgenomic Era

1. See Shiva (1997) and Hayden (2003) for excellent studies on how forms of biodiversity (both plants and animals as well as indigenous knowledge systems) have become important to the productive development of the biopharmaceutical, agricultural science, and bioscience industries.
2. The claim that the project of Western modern science has built into its epistemological framework a bifurcation between the knowable world (nature) and the unknowable world (social life) is one that science and technology theorists such as Bruno Latour and Donna Haraway have been arguing for the past 20 years. See Latour (1993) for perhaps the clearest account of this claim.
3. See Nicole Shukin (2009) for a very insightful analysis of the ways animals and their byproducts have become a part of a larger economic and cultural system of capitalizing on forms of life.
4. At the beginning of the 2010 school year, almost all nine winning state (and the District of Columbia) applications for Race to the Top Phase 2 funds emphasized STEM education (Robelen 2010). Indeed, the Obama Department of Education, behind its "educate to innovate" program, has designated STEM education as having "competitive preference priority" in the Race to the Top evaluation process, spending $3 billion annually on STEM education alone (Cavanagh 2008).
5. The method of teaching the controversy in the science classroom has multiple adherents. For example, the intelligent design movement largely led by the Discovery Institute also promotes a teaching the controversy method in the science classroom,

but for very different reasons than educators who are interested in making connections between science and society in their pedagogy. For intelligent design educators, it is a way in which to call into question evolutionary explanations for the origins of human life and open the way for creationist accounts. See Pierce (2007) for an examination of the intelligent design movement and how it unwittingly draws the teaching and learning of science into a productive political debate.

6. Pedretti and Nazir (2011) use the term STSE to designate the inclusion of "environment" to the STS movement in education. My own understanding does not make such a distinction since STS, at least in the figure of Latour, cannot be thought of without "nature" being a central concern, so in this sense it seems redundant. With that said, I also recognize the different disciplines in education such as environmental education and ecojustice education that overlap with STS, so the addition of environment makes sense in that it demarks a disciplinary convergence and shared concerns.
7. Pacific salmon belong to a different species *and* different genus than the Atlantic salmon (*Salmo salar*). The six species of Pacific salmon (Oncorhynchus) are Chinook (king), Coho (silver), Sockeye (red), Chum (dog), Pink (humpback or humpy), and Cherry. As will become apparent in my analysis, the species and genus boundaries of these basic biological classifications no longer hold up in the postgenomic age.
8. The American Society for Environmental History, for example, was created in 1977 and the journals *Environmental Review* and *Environmental History Review* published their first issues in 1976 and 1990, respectively. For two of the best examples of comprehensive and novel works in the field recently see both Ted Steinberg (2008) and Mark Fiege (2012).
9. See Coll Thrush (2007) for an excellent study of the way native iconography was used by city boosters and government officials to sell the Puget Sound region (specifically Seattle) as a utopian and "wild" land in an effort to settle the Northwest territory.
10. The Oregon country, the area of which Taylor is concerned in his study of the Pacific salmon fisheries crisis, spanned from Northern California (the Klamath river area) well into contemporary Southern British Columbia.
11. Daniel Boxberger (1989) also offers an insightful ethnohistory of the Lummi of the Pacific Northwest and their nonindustrial fishing and cultural practices.
12. See, for example, Gregory Cajete's excellent work on bridging Western approaches to teaching science with "ethnoscience" which he defines as "the study of a process by which every cultural group develops strategies to make nature accessible to reasoned enquiry. As a basic social thought process and way of adapting to a natural environment, it is unique to each culture, yet it reflects a common breeding ground between all cultures—the human mind" (1989, 73). Also see Brayboy and Maughan's (2009) work on the incorporation of indigenous knowledge into science lessons for teacher preparation programs.
13. Recombinant DNA (rDNA) is DNA that is artificially created through technoscientific means and used to combine two or more genetic sequences that would not normally occur together in nature (Haraway 1997).
14. If rooting enquiry in problems that face a community seems expressly Deweyan, there is good reason. ANT, and specifically the work of Latour, points to Dewey's work on "scientific democracy" in his *The Public and Its Problems* as a precursor model of politics for ANT to follow. In discussing the reformulation of politics along the lines of ANT, Latour states, "No one has made the point as forcefully

as John Dewey did with his own definition of the public. For a social science to become relevant, it has to have the capacity to renew itself—a quality impossible if a society is supposed to be 'behind' political action. It should also posses the ability to loop back from the few to the many and from the many to the few—a process often simplified under the terms of representation of the body politic" (Latour 2005, 261). Also for Dewey's proto-ANT pedagogy, see his discussion in *The School and Society* where he advocates for a "method [that] involves presentation of a large amount of detail, of minutiae of surroundings, tools, clothing, household utensils, foods, modes of living day by day, so that the child can reproduce the material of life, not as mere historic information. In this way, social processes and results become realities" (1990, 108).

15. See Richard White's (1995) excellent account of the damming of the Columbia river and its network effects on salmon.
16. Sandra Harding (2008) has argued in much of her recent work that part of forging and constructing more robust sciences will require drawing upon the marginalized epistemologies of women and "modernity's Others." That is, for Harding, "sciences from below" challenges the dominant androcentric and imperial model of Western science by offering a richer and more accurate account of the ways reality is interpreted from a variety of standpoints as opposed to the standard unitary one promoted throughout modernity.
17. For example, as David Gruenewald (2002) has pointed out in his excellent work on place-based education and Henry David Thoreau, experimentation and observation can also take place in transformative contexts outside the classroom where these terms become relational to the experience of place and what constitutes its pedagogical vitality.
18. The term actor or actant is used in ANT to designate "any entity that modifies another entity in a trial; of actors it can only be said that they act; their competence is deduced from their performances; the action, in turn, is always recorded in the course of a trial and by an experimentation protocol, elementary or not" (Latour 2004, 237).
19. One of the most important people who have brought to light the ecologically destructive "side effects" of Atlantic salmon aquafarming in the Northwest has been biologist and activist Alexandra Morton. Over the past 20 years, Morton has almost single-handedly held the aquafarming industry accountable for ecological problems that have spread from pen-raised fish. She was the first to discover how sea lice were being transferred from pen-raised salmon species to native populations, which caused a serious backlash within the Canadian government and the aquafarming industry. Recently, Morton has also been at the forefront of identifying a deadly virus (infectious salmon anemia ISA) in pen-raised salmon that has the potential to spread to native populations which the aquafarming industry in Canada denies. Perhaps most important to the topic of this study, Morton has done almost all of her work independent of institutional science, although she works closely with scientists in local British Columbia universities. In this sense, Morton is practicing science and democracy simultaneously while drawing on her scientific training in biology and research experience on Orca whales to address a public problem that is being ignored and actively resisted by both the aquafarming industry and the Canadian government. See Welch (2012) for an account of Morton's work on sea lice and now the ISA virus. Also, see her blog where she shares her research on salmon and the aquafarming industry online at: http://alexandramorton.typepad.com/

20. Over the past ten years, mounting scientific evidence has shown the devastating ecological impact pen-raised Atlantic salmon have on Northwest migratory waterways used by wild Pacific salmon. The single largest problem stems from the transfer of sea lice from pen-raised fish to wild salmon fry which cannot withstand the parasite in their vulnerable state of development. Pens suspended in water where rivers and streams meet the sea is one of the primary growing methods used by auqafarming corporations for producing fish. See, for example, Krkosek et al. (2007) and Hindar et al. (2006).

5 The Biomedicalization of Kids: Psychotropic Drugs and Biochemical Governing in High Stakes Schooling

1. In the rather extensive educational literature on Foucault, there have been a number of studies on the effect of a medical gaze on students' bodies. See Berglund (2008) and Lewis' (2010) work that, respectively, demonstrate how the medicalization of bodies is an important facet to current expressions of educational biopower. For Foucauldian perspectives in education that look specifically at how new technologies are helping form novel forms of educational governmentality in relation to students' bodies, see Wong (2009) and Nadesan (2009) where brain-based learning and autism are utilized as anchoring points to better understand the convergence of technologies of self and educational governing.
2. While I only focus on the use of psychotropic drugs in K-12 schooling, over the past few years there has been an explosion in black market usage of "cognitive-enhancing" drugs such as Ritalin and Adderall on college campuses across the United States. In a celebratory account of this growth trend in psychotropic drug use in higher education settings by students to enhance their academic performance, scientists in a recent *Nature* article point out that "almost 7% of students in US universities have used prescription stimulants in this way, and that on some campuses, up to 25% of students had used them in the past year. These students are early adopters of a trend that is likely to grow, and indications suggest that they're not alone" (Greely et al. 2008, 702). For the scientists who wrote this article, at least one of whom works for the pharmaceutical industry, the responsible use of cognitive-enhancing drugs by the healthy is no different from other technologies we use to increase our cognitive production abilities such as education, nutrition, prosthetic brain chips, and sleep.
3. In addition to the rapid growth in ADHD diagnoses and treatment of children in the United States, a corollary pattern in bipolar diagnoses in children has also emerged. While the connection between ADHD and bipolar diagnoses in children is beyond the scope of this chapter, it is important to keep in mind that one reason the two "disorders" are connected is due to the biochemical cocktails created through the mixing of psychotropic drugs to treat side effects that arise with each new medication introduced into the body of children. For example, one of the major side effects of ADHD drugs such as Ritalin and Adderall is depression in children. Depression caused by ADHD drugs, in turn, leads to the prescription of antidepressant drugs such as Wellbutrin which are added to the treatment plan of children, which, as a result, may cause high anxiety and thus a prescription for an antianxiety drug such as Xanax is added to the mix and so on. Such mixing of multiple drugs at once in an attempt to cancel out each others' side effects has lead to serious neurological damage such as uncontrollable ticks as well as suicidal

behavior. See PBS's *Frontline: The Medicated Child* (2008) for an in-depth and well-researched examination of the explosion in bipolar diagnoses in children and the adverse effects of drug cocktails.

4. In the marketing of drugs, pharmaceutical companies often test and develop names for their products that elicit a quality the customer is searching for in terms of treatment. Viagra, for example, is quite clearly aiming to target those seeking vitality in their sexual lives, while Concerta suggests a heightened intellectual ability to concentrate or to stay on task. There is little doubt that such a name was market-tested and won out over other potential brand titles to exude the precise qualities that individuals (or parents) wanting to enhance concentration abilities would seek.
5. See Henry Giroux's (2004) excellent work on developing the concept of public pedagogy within the neoliberal context of education and in particular, how the media is central to instantiating free-market fundamentalist values within the individual.
6. In 2009, this figure has risen to over $300 billion/year, achieving a growth rate of 5.1 percent over a 1.8 percent growth rate in 2008 (Berkrot 2010).
7. In the mid-1990s, PBS aired a Merrow Report story *ADD: A Dubious Diagnosis?*" which brought the CHADD scandal into public discussion for the first time. See Breggin (2001) for a full account of the repercussions of the story and how CHADD and the pharmaceutical industry had to become less obvious partners. For a more recent account of CHADD's more subtle ties to pharmaceutical corporations also, see Fitzgerald (2009).
8. Also see Nikolas Rose's (1999) chapter "The Child, the Family, and the Outside World" for a strong analysis of the ways psychiatric knowledge intervened into the family structure of Western cultures through a focus on mothers.
9. The term Ritalin wars dubbed by the media was meant to signify the intense controversy that emerged in the 1990s in the United States between advocates of using psychostimulant drugs to treat behavioral "disorders" in children and those against their pervasive use.
10. See the "Chronicle of Higher Education" article on Dr. Nemeroff's ties to Insel and the NIMH, who also was forced to leave his editor position at the journal of *Neuropsychopharmacology* in 2006 because "he was reported to have endorsed an implantable device for treating depression without disclosing payments from its manufacturer" (Basken 2010). Nemeroff has also forced the prestigious journal *Nature* to change its policies for authors' disclosures after he praised treatments for depression in which he had an unreported financial interest.
11. It should also be noted that in 1995 former NIH director Harold Varmus instituted a policy that "placed no limits on the amount of money its scientists could earn from outside work or the time they devote to it" (Angell 2004, 104).
12. See Breggin's (2001) point-by-point critical analysis of the NIMH study that, as he points out, fails to meet basic methodological and scientific standards.
13. See Breggin's (2001) analysis of the most well-publicized NIMH study that utilized brain scanning technology to validate a biological basis for ADHD.
14. Barkley has published numerous popular books that present themselves as educational tools for parents and families dealing with ADHD children or potential cases. His *Attention-Deficit Hyperactivity Disorder: A Handbook for Diagnosis and Treatment* (1998) and *Taking Charge of ADHD: The Complete, Authoritative Guide for Parents* (2000) are two of the best-selling ADHD self-help titles on the market. For someone who claims that ADHD is 80 percent a matter of hereditability

(Diller, 1998) and that psychostimulant drugs achieve a 70 percent improvement rate (Breggin 2001), it is not surprising that Barkley pushes drug intervention as the most effective method for increasing children's learning abilities and overall mental health.

15. Resistance to neoliberal development policies that promote genetic appropriation from indigenous peoples and nonhumans (plants, for example) has expressed itself in a multitudinal form. That is, many indigenous groups across the globe from New Zealand, First Nation's People in Canada, American Indian groups in the United States, and India have been leading movements against forced developmental plans and the "plundering of nature and knowledge" associated with the needs of biocapitalist growth. For an excellent documentary examining how new genetic sciences allow for a whole new level of colonial extraction of value, see the Indigenous Peoples Council on Biocolonialism's film *The Leech & the Earthworm*.
16. The food sovereignty movement, as researchers and activists such as Vanda Shiva and Raj Patel have documented extensively in their work, is a broad based yet situated resistance movement by indigenous groups working in alliance with other groups across multiple continents. Most all food sovereignty movements share the view that the further privatization and colonial appropriation of biological materials *and* indigenous knowledge systems and practices represent a new phase of colonial enclosure. For a concise and powerful statement on biotechnologies and indigenous peoples, see Tauli-Corpuz (2001) as well as Shiva (1997, 2005) and Patel (2008, 2009) for a more general treatment of the food sovereignty movement and its relation to free-market development policies.

Epilogue

1. See Gregory N. Bourassa's (2011) excellent critical analysis of the literature on biopolitics in educational theory.
2. Richard Kahn (2010), in his development of a theory and practice of ecopedagogy, has also argued that technique needs to be reconstructed in a sustainable and interspecies manner. Kahn also offers a strong history of theorists and activists (such as Herbert Marcuse, Ivan Illich, and Judi Bari) who have centered the demand for taking science and technology out of the productive engine of capital and placing it instead in the service of autonomous, sustainable, and just communities.
3. See the International Science Shop Network website where they share projects they are working and particularly how scientific research is cast in a co-productive framework with the public. Online at: http://www.livingknowledge.org/livingknowledge/

Bibliography

Aikenhead, G. *Science Education for Everyday Life: Evidence-based Practice.* New York: Teachers College Press, 2006.

———. "What Is STS in Science Teaching?" In *STS Education: International Perspectives on Reform*, edited by J. Solomon and G. Aikenhead. New York: Teachers College Press, 1994.

Aikenhead, G., and J. Solomon. *STS Education: International Perspectives on Reform.* New York: Teachers College Press, 1994.

Akinbami, L. J., X. Liu, P. N. Pastor, and C. A. Reuben. "Attention Deficit Hyperactivity Disorder among Children Aged 5–17 Years in the United States, 1998–2009." *Center for Disease Control and Prevention.* NCHS Data Brief 70. 2011.

Allman, P. *Critical Education against Global Capitalism: Karl Marx and Revolutionary Critical Education.* Westport, CT: Bergin & Garvey, 2001.

American Recovery and Reinvestment Act. 2009. Retrieved November 12, 2011, from *appropriations.house.gov/pdf/RecoveryBill01–15–09.pdf.*

Angell, M. *The Truth about the Drug Companies: How They Deceive Us and What to Do About It.* New York: Random House, 2004.

AquaBounty Technologies. "AquaCulture Market." 2010. Retrieved May 23, 2011, from http://ww.aquabounty.com/company/aquaculture-293.aspx.

Association of American Universities. "National Defense Education and Innovation Initiative: Meeting America's Economic and Security Challenges in the 21st Century." 2006. Retrieved October 17, 2011, from www.aau.edu/reports/NDEII.pdf.

Baichwal, J. (dir.) 2007. *Manufactured Landscapes.* Zeitgeist Films DVD.

Ball, S. J. *Foucault and Education: Discipline and Knowledge.* New York: Routledge, 1990.

Ball, S. J., and C. Junemann. *Networks, New Governance and Education.* Bristol, UK: Policy Press, 2012.

Barkley, R. 1998. *Attention-Deficit Hyperactivity Disorder: A Handbook for Diagnosis and Treatment.* New York: Guilford Press, 2012.

———. *Taking Charge of ADHD: The Complete, Authoritative Guide for Parents.* New York: Guilford Press, 2000.

Basken, P. "As He Worked to Strengthen Ethics Rules, NIMH Director Aided a Leading Transgressor." *The Chronicle of Higher Education.* 2010. Retrieved April 2, 2012, from http://chronicle.com/article/While-Revising-Ethics-Rules/65800/.

Becker, G. *Human Capital: A Theoretical and Empirical Analysis, with Special Reference to Education.* Chicago: University of Chicago Press, 1994 [1964].

Berglund, G. "Pathologizing and Medicalizing Lifelong Learning: A Deconstruction." In *Foucault and Lifelong Learning: Governing the Subject*, edited by A. Fejes and K. Nicoll, 138–150. New York: Routledge, 2008.

Berkrot, B. "U.S. Prescription Drug Sales Hit $300 Billion in 2009." 2010. Retrieved April 11, 2012, from http://www.reuters.com/article/2010/04/01/us-drug-sales-idUSTRE6303CU20100401.

Best, S. and D. Kellner. *The Postmodern Adventure: Science, Technology, and Cultural Studies at the Third Millennium*. New York: Guilford Press, 2001.

Boger, J. C. "Education's 'Perfect Storm'? Racial Resegregation, "High-Stakes Testing", & School Inequities: The Case of North Carolina." *The Civil Rights Project/Proyecto Derechos Civiles*. 2002. Retrieved November 23, 2012, from http://civilrightsproject.ucla.edu/.

Borón, A. *Empire and Imperialism: A Critical Reading of Michael Hardt and Antonio Negri*. London: Zed Books, 2005.

Bourassa, G. N. "Rethinking the Curricular Imagination: Curriculum and Biopolitics in the Age of Neoliberalism." *Curriculum Inquiry* 41.1 (2011): 5–16.

Bowles, S., and H. Gintis. *Schooling in Capitalist America: Educational Reform and the Contradictions of Economic Life*. New York: Basic Books, 1977.

———. "The Problem with Human Capital Theory—A Marxian Critique." *American Economic Review* 63.2 (1975): 74–82.

Boxberger, D. *To Fish in Common: The Ethnohistory of Lummi Indian Salmon Fishing*. Seattle, WA: University of Washington Press, 1989.

Brayboy, B. M. J., and E. Maughan. "Indigenous Knowledges and the Story of the Bean." *Harvard Educational Review* 79.1 (2009): 1–21.

Breggin, Peter. R. *Talking Back to Ritalin: What Doctors Aren't Telling You about Stimulants and ADHD*. Cambridge, MA: De Capo Press, 2001.

Breggin, Peter. R., and G. R. Breggin. *The War against Children of Color: Psychiatry Targets Inner-City Youth*. Monroe, Maine: Common Courage Press, 1998.

Brickhouse, N. "Teaching Sciences: The Multicultural Question Revisited." *Science Education* 85.1 (2003): 35–49.

Brickhouse, N., and J. Kittleson. "Visions of Curriculum, Community and Science." *Educational Theory* 56.2 (2006): 191–204.

Briggs, D. and Domingue, B. 2011. Due Diligence and the Evaluation of Teachers: A review of the value-added analysis underlying the effectiveness rankings of Los Angeles Unified School District teachers by the *Los Angeles Times*. National Education Policy Center. http://nepc.colorado.edu/publication/due-diligence.

Bröckling, U. "Human Economy, Human Capital: A Critique of Biopolitical Economy." In *Governmentality: Current Issues and Future Challenges*, edited by U. Bröckling, S. Krasmann, and T. Lemke. New York: Routledge, 2011.

Bröckling, U., S. Krasmann, and T. Lemke. "From Foucault's Lectures at the Collège de France to Studies of Governmentality." In *Governmentality: Current Issues and Future Challenges*, edited by U. Bröckling, S. Krasmann, and T. Lemke. New York: Routledge, 2011a.

———, eds. *Governmentality: Current Issues and Future Challenges*. New York: Routledge, 2011b.

Bullard, R. D. *Dumping in Dixie: Race, Class, and Environmental Quality*. Boulder, CO: Westview Press, 2000.

Cajete, G. *Igniting the Sparkle: An Indigenous Science Education Model*. Skyland, NC: Kivaki Press, 1999.

Camfield, D. "The Multitude and the Kangaroo: A critique of Hardt and Negri's Theory of Immaterial Labour." *Historical Materialism* 15.2 (2007): 21–52.

Cavanagh, S. "Federal Projects' Impact on STEM Remains Unclear." *Education Week.* 2008. Retrieved December 6, 2011, from http://www.edweek.org/login.html?source=http://www.edweek.org/ew/articles/2008/03/27/30.

Center for Disease Control and Prevention (CDC). 2008. Diagnosed Attention Deficit Hyperactivity Disorder and Learning Disability: United States, 2004–2006. *Vital and Health Statistics* 10.237@@

———. 2010. Increasing Prevalence of Parent-Reported Attention-Deficit /Hyperactivity Disorder Among Children—United States, 2003 and 2007. *Morbidity and Mortality Weekly Report.* http://www.cdc.gov/mmwr/preview/mmwrhtml/mm5944a3.htm?s_cid=mm5944a3_w

———. 2011. Rates of Parent-Reported ADHD Increasing. http://www.cdc.gov/features/dsADHD/

Chambers, J. "Michigan orders DPS to Make Huge Cuts." *The Detroit News.* 2011. Retrieved January 8, 2012, from http://www.detnews.com/article/20110221/SCHOOLS/102210355/1409/Michigan-orders-DPS-to-make-huge-cuts.

Clowes, G. A. "Helping Teachers Raise Student Achievement: An Interview with William L. Sanders." *The Heartland Institute.* 1999. Retrieved October 23, 2011, from http://www.heartland.org/policybot/results/11119/Helping_Teachers_Raise_Student_Achievement_an_interview_with_William_L_Sanders.html.

Conrad, A. H., and J. R. Meyer. "The Economics of Slavery in the Ante Bellum South." *Journal of Political Economy* 66.2 (1958): 95–130.

Coon, C. S. *The Origin of Races.* New York, NY: Alfred A. Knopf, 1962.

Cooper, M. *Life as Surplus: Biotechnology and Capitalism in the Neoliberal Era.* Seattle, WA: University of Washington Press, 2008.

Cooper, M. H. "Commercialization of the University and Problem Choice by Academic Biological Scientists." *Science, Technology, & Human Values* 34.5 (2009): 629–653.

Council for Biotechnology Information (CBI). *Look Closer at Biotechnology.* 2012. Retrieved May 21, 2012, from http://www.whybiotech.com/resources/activity-book.asp.

Cronon, W. *Nature's Metropolis: Chicago and the Great West.* New York: W.W. Norton & Company, 1991.

———. "The Trouble with Wilderness; or Getting Back to the Wrong Nature." In *Uncommon ground: Rethinking the Human Place in Nature,* edited by W. Cronon. New York: W.W. Norton & Company, 1995.

Crosby, A. *Ecological imperialism: The Biological Expansion of Europe, 900–1900.* New York, NY: Cambridge University Press, 1986.

———. *The Columbian Exchange: Biological and Cultural Consequences of 1492.* Westport, CT: Praeger Publishers, 2003 [1973].

Crowe, C. "Time on the Cross: The Historical Monograph as a Pop Event." *History Teacher* 9.4 (1976): 588–630.

Darling-Hammond, L. *Creating a Comprehensive System for Evaluating and Supporting Effective Teaching.* Stanford, CA: Stanford Center for Opportunity Policy in Education, 2012.

———. *The Flat World and Education: How America's Commitment to Equity Will Determine Our Future.* New York: Teacher's College Press, 2010.

Darling-Hammond, L., A. Amrein-Beardsley, E. Haertel, and J. Rothstein. "Evaluating Teacher Evaluation." *Phi Delta Kappan* 93.6 (2012): 8–15.

Davidson-Harden, A. "Neoliberalism, Knowledge Capitalism and the Steered University: The role of OECD and Canadian Federal Government Discourse." In *Governmentality Studies in Education,* edited by M. Peters, A. C. Besley, M. Olssen, S. Maurer, and S. Weber. Rotterdam, Amsterdam: Sense Publishers, 2009.

Davis, A. Y. *Abolition Democracy: Beyond Prisons, Torture, and Empire Interviews with Angela Y. Davis*. New York: Seven Stories Press, 2005.

———. *Women, Race, and Class*. New York: Vintage Books, 1983.

Dean, M. *Governmentality: Power and Rule in Modern Society*. Los Angeles, CA: Sage, 2010.

Delgado Bernal, D. "Critical Race Rheory, LatCrit Theory and Critical Raced-Gendered Epistemologies: Recognizing Students of Color as Holders and Creators of Knowledge." *Qualitative Inquiry* 8.1 (2002): 1005–126.

Department of Education (DOE). *Identifying and Treating Attention Deficit Hyperactivity Disorder: A Resource for School and Home*. Jessup, MD: ED Pubs, Education Publication Center, 2003.

———. *Teaching Children with Attention Deficit Hyperactivity Disorder: Instructional Strategies and Practices*. Jessup, MD: ED Pubs, Education Publication Center, 2008.

Deutsch, M. "Happenings on the Way Back to the Forum: Social Science, IQ, and Race Differences Revisited." *Harvard Educational Review* 39.3 (1969): 523–557.

Dewey, J. *The Public and Its Problems*. Athens, Ohio: Ohio University Press, 1954.

———. *The School and Society*. Chicago: University of Chicago Press, 1990 [1900].

Diller, L. H. *Running on Ritalin: A Physician Reflects on Children, Society, and Performance in a Pill*. New York: Bantam Books, 1998.

Diller, L.H. "What Could—And Couldn't—Be Causing America's ADHD Epidemic." 2011. Retrieved March 24, 2012, from http://www.huffingtonpost.com/larry-diller/united-states-of-adderall_b_980238.html.

District of Columbia Public Schools. "What is Value-Added?" 2011. Retrieved September 7, 2012, from http://www.dc.gov/DCPS/In+the+Classroom/Ensuring+Teacher+Success/IMPACT+%28Performance+Assessment%29/Value-Added.

Du Bois, W. E. B. *Black Reconstruction in America: 1860–1880*. New York: Free Press, 1999 [1935].

———. *Darkwater: Voices from within the Veil*. Atlanta, GA: Supreme Design Publishing, 2011.

———. *Dusk of Dawn: An Essay toward an Autobiography of a Race Concept*. New Brunswick, NJ: Transaction Publishers, 1995.

———. "Education and Work." In *Du Bois on Education*, edited by Eugene Provenzo Jr. Walnut Creek, CA: AltaMira Press, 1930.

———. "Negros in College." In *Du Bois on Education*, edited by Eugene Provenzo Jr. Walnut Creek, CA: AltaMira Press, 1926.

———. *The Education of Black People: Ten Critiques, 1906–1960*. New York: Monthly Review Press, 2001.

———. *The Gift of Black Folk: The Negroes and the Making of America*. New York: Square One Publishers, 2009.

———. *The Souls of Black Folk*. New York: Oxford University Press, 2007.

DuPaul, G. J., and G. Stoner. *ADHD in the Schools: Assessment and Intervention Strategies*. New York: Guilford Press, 2003.

Durant, M. W. *Economic Value Added: The Invisible Hand at Work*. Columbia, MD: Credit Research Foundation, 1999.

Dyer-Witherford, N. "Empire, Immaterial Labor, the New Combinations, and the Global Worker." *Rethinking Marxism Rethinking Marxism: A Journal of Economics, Culture, & Society* 13.3 (2001): 70–80.

Engels, F. *The Origin of the Family, Private Property and the State*. New York: Penguin, 2010.

Esposito, R. *Bios: Biopolitics and philosophy.* Minneapolis, MN: University of Minnesota Press, 2008.

Fassin, D. "Another Politics of Life Is Possible." *Theory, Culture & Society* 26.44 (2009): 44–61.

Felch, J. and J. Song. "U.S. Schools Chief Endorses Release of Teacher Data." *The Los Angeles Times.* 2010. Retrieved October 29, 2011, from http://articles.latimes.com/2010/aug/16/local/la-me-0817-teachers-react-20100817.

Felch, J., J. Song, and D. Smith. "Who's teaching L.A.'s kids?" *The Los Angeles Times.* 2010. Retrieved September 21, 2011, from http://www.latimes.com/news/local/la-me-teachers-value-20100815,0,2695044.story.

Fiege, M. *The Republic of Nature: An Environmental History of the United States.* Seattle, WA: University of Washington Press, 2012.

Fine, M., A. Burns, Y. Payne, and M. Torre. Civic Lessons: The Color and Class of Betrayal. *Teachers College Record* 106.11 (2004): 2193–2223.

Fitzgerald, T. D. *White Prescriptions?: The Dangerous Social Potential for Ritalin and Other Psychotropic Drugs to Harm Black Boys.* Boulder, CO: Paradigm Publishers, 2009.

Fogel. R. W. *The Escape from Hunger and Premature Death, 1700–2100.* New York: Cambridge University Press, 2004.

———. *Without Consent or Contract: The Rise and Fall of American Slavery.* New York: W. W. Norton & Company, 1989.

Fogel, R. W., and Stanley L. Engerman. *Time on the Cross: The Economics of American Negro Slavery.* Boston: Little, Brown and Company, 1974.

Fogel, R. W., R. Floud, B. Harris, and S. C. Hong. *The Changing Body: Health, Nutrition, and Human Development in the Western World since 1700.* New York: Cambridge University Press, 2011.

Food and Drug Administration Center for Veterinary Medicine (FDACVM). *Briefing Packet for AquAdvantage Salmon.* 2010. Retrieved November 9, 2011, from www.fda.gov/downloads/AdvisoryCommittees/ ... /UCM224762.pdf.

Forti, S. "The Biopolitics of Souls: Racism, Nazism and Plato." *Political Theory* 34.9 (2006): 9–32.

Foucault, M. *Society Must be Defended: Lectures at the Collège de France 1975–1976.* Translated by D. Macey. New York: Picador, 2003.

———. *The Birth of Biopolitics: Lectures at the Collège de France1978–1979.* Translated by G. Burchell. New York: Palgrave Macmillan, 2008.

———. *The History of Sexuality Volume 1: An Introduction.* New York: Vintage Books, 1978.

Franklin, S., and S. McKinnon, eds. *Relative Values: Reconfiguring Kinship Studies.* Durham, NC: Duke University Press, 2001.

Franklin, S., and M. Lock. "Animation and Cessation: The remaking of life and death." In *Remaking Life and Death: Toward an Anthropology of the Biosciences,* edited by S. Franklin and M Lock. Santa Fe, NM: School of American Research Press, 2003.

Franklin, S., and M. Lock, eds. *Remaking Life & Death: Toward an Anthropology of the Biosciences.* Santa Fe, New Mexico: School of American Research Press, 2003.

Freire, P. *Pedagogy of the Oppressed.* New York: Continuum, 2000 [1970].

Friedman, M. "Public Schools: Make Them Private." *Education Economics* 5.3 (1995): 341–344.

Friedman, T. *Hot, Flat, and Crowded: Why We Need a Green Revolution – -and How It Can Renew America.* New York: Picador, 2009.

———. *The World is Flat 3.0: A Brief History of the Twenty-First Century.* New York: Picador, 2007.

Gagon, M. A., and J. Lexchin. 2008. The Cost of Pushing Pills: A new Estimate of Pharmaceutical Promotion Expenditures in the United States. *PLoS Med* 5.1. http://www.plosmedicine.org/article/info:doi/10.1371/journal.pmed.0050001

Gates, B. "Shame Is Not the Solution." *The New York Times.* 2012. Retrieved May 16, 2012, from http://www.nytimes.com/2012/02/23/opinion/for-teachers-shame-is-no-solution.html

Gaviria, M. 2008. (dir.) *Frontline: The Medicated Child.* PBS.

Gilroy, P. *The Black Atlantic: Modernity and Double-Consciousness.* Cambridge: Harvard University Press, 1993.

Gingrich, N. *Winning the Future: A 21st Century with America.* Washington DC: Regnery Publishing, 2006.

Giroux, H. *Against the Terror of Neoliberalism: Politics Beyond the Age of Greed.* Boulder, CO: Paradigm Publishers, 2008.

———. "Cultural Studies of a Public Pedagogy: Making the Political More Pedagogical." *Parallax* 10. 2 (2004): 73–89.

———. *Disposable Youth: Racialized Memories, and the Culture of Cruelty.* New York: Routledge, 2012.

———. *Youth in a Suspect Society: Democracy or Disposability?* New York: Palgrave Macmillan, 2010.

Goldin, C. "Cliometrics and the Nobel." *Journal of Economic Perspectives* 9.2 (1995): 191–208.

Goldin, C., and L. F. Katz. *The Race between Education and Technology.* Cambridge, MA: Harvard University Press, 2010.

Gonzales, N., L. C. Moll, M. F. Tenery, A. Rivera, P. Rendon, R. Gonzales, and C. Amanti. "Funds of Knowledge for Teaching in Latino Households." *Urban Education* 29.4 (1995): 443–470.

Gordon, C. "Governmental Rationality: An Introducation." In *The Foucault Effect: Studies in Governmentality*, edited by G. Burchell, C. Gordon, and P. Miller. Chicago, IL: University of Chicago Press, 1991.

Greely, H., B. Sahakian, J. Harris, R. C. Kessler, M. Gazzaniga, P. Campbell, and M. J. Farah. "Towards Responsible Use of Cognitive-Enhancing Drugs by the Healthy." *Nature* 456, 2008.

Greven, U. C., N. Harlaar, Y. Kovas, T. Chamorro-Premuzic, and R. Plomin. "More than Just IQ: School Achievement Is Predicted by Self-Perceived Abilities—But for Genetic Rather than Environmental Reasons." *Psychological Science* 20.6 (2009): 753–762.

Gruenewald, D. Teaching and Learning with Thoreau: Honoring Critique, Experimentation, Wholeness, and the Places Where we Live. *Harvard Educational Review* 72.4 (2002): 515–541.

Guggenheim, D. *Waiting for "Superman".* Los Angeles, CA: Walden Media, 2010.

Gutman, H. G. *Slavery and the Numbers Game: A Critique of Time on the Cross.* Urbana-Champaign, IL: University of Illinois Press, 1975.

Haraway, D. *Modest_Mouse@Second_Millennium.Female©_Meets_OncoMouse™.* New York: Routledge, 1997.

———. *Simians, Cyborgs, and Women: The Reinvention of Nature.* New York: Routledge, 1990.

———. *When Species Meet.* Minneapolis, MN: University of Minnesota Press, 2008.

Harding, S. *Sciences from Below: Feminisms, Postcolonialities, and Modernities.* Durham, NC: Duke University Press, 2008.

———, ed. *The Feminist Standpoint Theory Reader: Intellectual and Political Controversies*. New York: Routledge, 2003.

Hardt, M., and A. Negri. *Commonwealth*. Cambridge: Harvard University Press, 2009.

———. *Empire*. Cambridge, MA: Harvard University Press, 2000.

———. *Multitude: War and Democracy in the Age of Empire*. New York, NY: Penguin Press, 2004.

Harris, N. D. *Value-Added Measures in Education: What Every Educator Needs to Know*. Cambridge, MA: Harvard Education Press, 2011.

Harvey, D. *A Brief History of Neoliberalism*. New York: Oxford University Press, 2005.

———. *The Enigma of Capital and the Crisis of Capitalism*. New York: Oxford University Press, 2010.

Hayden, C. *When Nature Goes Public: The Making and Unmaking of Bioprospecting in Mexico*. Princeton, NJ: Princeton University Press, 2003.

Headrick, D. *The Tools of Empire: Technology and European Imperialism in the Nineteenth Century*. Oxford: Oxford University Press, 1981.

Helmreich, S. "Species of Biocapital." *Science as Culture* 17.4 (2008): 463–478.

Herrnstein, R. J., and C. Murray. *The Bell Curve: Intelligence and Class Structure in American Life*. New York: Free Press, 1996.

Hershberg, T. "Value-Added Assessment and Systemic Reform: A Response to America's Human Capital Development Challenge." 2005. www.cgp.upenn.edu/pdf/aspen.pdf.

Hindar, K., I. A. Fleming, P. McGinnity, and O. Diserud. "Genetic and Ecological Effects of Salmon Farming on Wild Salmon: Modeling from Experimental results. *Journal of Marine Science* 63 (2006): 1234–1247.

Hines, P. J. Why Controversy Belongs in the Science Classroom: From Bioengineered Food to Global Warming, Science is Rife with Dispute, Debate, and Ambiguity." *Harvard Educational Letter* 17.5 (2001): 7–8.

Hodson, D. "Time for Action: Science Education for an Alternative Future." *International Journal of Science Education* 25.6 (2003): 645–670.

———. *Towards Scientific Literacy: A Teacher's Guide to the History, Philosophy and Sociology of Science*. Rotterdam, Netherlands: Sense Publishers, 2008.

Hopkinson, N. "Why School Choice Fails." *The New York Times*. 2011. http://www.nytimes.com/2011/12/05/opinion/why-school-choice-fails.html?_r=1.

Horkheimer, M., and T. Adorno. *Dialectic of Enlightenment*. Translated by Edmund Jephcott. Palo Alto, CA: Stanford University Press, 2001 [1947].

Ignatiev, N. *How the Irish Became White*. New York: Routledge, 2008.

Illich, I. *Shadow Work*. Salem, NH: Marion Boyars, 1981.

Institute of Bioengineering and Nanotechnology (IBN). 2011. http://www.ibn.a-star.edu.sg/about_ibn_0.php

Jensen, A. R. "How Much Can We Boost IQ and Scholastic Achievement?" *Harvard Educational Review* 39.1 (1969): 1–123.

Kahn, R. *Critical Pedagogy, Ecoliteracy, and Planetary Crisis: The Ecopedagogy Movement*. New York: Peter Lang, 2010.

Kantor, H., and R. Lowe. "The Price of Human Capital: Educational Reform and the Illusion of Equal Opportunity." *Dissent* (Summer 2011).

Kawagley, A. O., D. Norris-Tull, and R. A. Norris-Tull. "The Indigenous Worldview of Yupiaq Culture: Its Scientific Nature and Relevance to the Practice and Teaching of Science." *Journal of Research in Science Teaching* 35.2 (1998): 133–144.

Kellner, D. *Critical Theory, Marxism, and Modernity*. Baltimore, MD: Johns Hopkins University Press, 1989.

Kellner, D., and C. Pierce, eds. *Herbert Marcuse: Philosophy, Psychoanalysis, and Emancipation*. New York and London: Routledge, 2011.

King's College of London's Institute of Psychiatry, Social, Genetic & Developmental Psychiatry Centre. "About the Social, Genetic & Developmental Psychiatry Centre." 2012. http://www.kcl.ac.uk/iop/depts/mrc/about/index.aspx.

Kirkham, C. "Buying Legitimacy: How a Group of California Executives Built an Online College Empire." 2011. Retrieved February 7, 2012, from http://www.huffingtonpost.com/2011/03/09/ashford-university-for-profit-college_n_833735.html.

Klein, N. *The Shock Doctrine: The Rise of Disaster Capitalism*. New York: Henry Holt and Co., 2007.

Kleinman, D., and S. Vallas. "Science, Capitalism, and the Rise of the 'Knowledge Worker': The Changing Structure of Knowledge Production in the United States." *Theory and Society* 30.4 (2001): 451–492.

Krkosek, M., J. S. Ford, A. Morton, S. Lele, R. A. Myers, and M. A. Lewis. "Declining Wild Salmon Populations in Relation to Parasites from Farm Salmon." *Science* 318 (2007): 1772–1775.

Labaree, D. F. "Targeting Teachers." *Dissent* (Summer 2011): 9–14.

Ladson-Billings, G. *The Dreamkeepers: Successful Teachers of African American Children*. San Francisco, CA: Jossey-Bass, 2009.

LaDuke, W. "The Salmon People: Sussana Santos." In *The Winona LaDuke Reader*, edited by W. LaDuke. Stillwater, MN: Voyageur Press, 2002a.

———. *The Winona LaDuke Reader*. Stillwater, MN: Voyageur Press, 2002b.

Latour, B. *Politics of Nature: How to Bring the Sciences into Democracy*. Translated by C. Porter. Cambridge, MA: Harvard University Press, 2004.

———. *Reassembling the Social: An Introduction to Actor-Network-Theory*. New York: Oxford University Press, 2005.

———. *Science in Action: How to Follow Scientists and Engineers through Society*. Cambridge, MA: Harvard University Press, 1987.

———. *We Have Never Been Modern*. Translated by C. Porter. Cambridge, MA: Harvard University Press, 1993.

Laveaga, G. S. *Jungle Laboratories: Mexican Peasants, National Projects, and the Making of the Pill*. Durham, NC: Duke University Press, 2009.

Lazzarato, M. "Immaterial Labor" In *Radical Thought in Italy: A Potential Politics*, edited by Michael Hardt and Paolo Virno, 133–147. Minneapolis, MN: University of Minnesota Press, 1996.

Lemke, T. *Biopolitics: An Advanced Introduction*. Translated by Eric Frederick Trump. New York, NY: New York University Press, 2011.

———. "Foucault, Governmentality, and Critique." *Rethinking Marxism: A Journal of Economics, Culture, & Society* 14.3 (2002): 49–64.

———. " 'The Birth of Bio-politics': Michel Foucault's Lecture at the Collège de France on Neo-liberal Governmentality." *Economy and Society* 30.2 (2001): 190–207.

Lewis, D. L. *W.E.B. Du Bois: Biography of a Race, 1868–1919*. New York: Henry Holt & Co., 1993.

Lewis, T. E. "Biopolitical Utopianism in Educational Theory." *Educational Philosophy and Theory* 39.7 (2007): 683–702.

———. "Defining the Political Ontology of the Classroom: Toward a Multitudinous Education." *Teaching Education* 19.4 (2008): 249–260.

———. "Discipline-Sovereignty-Education." *A Foucault for the 21st Century: Governmentality, Biopolitics and Discipline in the New Millennium*, edited by S. Blinkley and J. Capetillo-Ponce, 173–187 Tyne, UK: Cambridge Scholars Publishing, 2010.

———. "Education and the Immunization Paradigm." *Studies in Philosophy and Education* 28.6 (2009): 485–498.

———. "The School as an Exceptional Space: Rethinking Education from the Perspective of the Biopedagogical." *Educational Theory* 56.2 (2006): 159–176.

Lewis, T., and R. Kahn. *Education out of Bounds: Reimagining Cultural Studies for a Posthuman Age.* New York, NY: Palgrave Macmillan, 2010.

Lichatowich, J. *Salmon without Rivers: A History of the Pacific Salmon Crisis.* Washington DC: Island Press, 1999.

Linder, C., L. Östman, D. A. Roberts, P. Wickman, G. Erickson, and A MacKinnon, eds. *Exploring the Landscape of Scientific Literacy.* New York, NY: Routledge, 2011.

Lipman, P. "Making the Global City, Making Inequality: The Political Economy and Cultural Politics of Chicago School Policy." *American Educational Research Journal* 39.2 (2002): 379–419.

Lukács, G. *The Destruction of Reason.* Translated by Peter Palmer. London: Merlin Press, 1980.

MacKenzie, D., and J. Wajcman, eds. *The Social Shaping of Technology.* Philadelphia, PA: Open University Press, 1999 [1985].

Mann, C. C. *1491: New Revelations of the Americas Before Columbus.* New York: Random House, 2006.

Marcuse, H. "Children of Prometheus: 25 Theses on Technology and Society." In *Herbert Marcuse: Philosophy, Psychoanalysis, and Emancipation*, edited by D. Kellner and C. Pierce, 222–225. New York and London: Routledge, 2011.

———. *One-Dimensional Man: Studies in the Ideology of Advanced Industrial Society.* Boston: Beacon Press, 1964 [1991].

Marshall, J. D. "Foucault and Neo-liberalism: Biopower and Busno-Power." In *Philosophy of Education Yearbook*, edited by A. Neiman, 320–329. Urbana-Champaign, Ill: Philosophy of Education Society,1995.

Marx, K. *Capital: Volume I.* Translated by Ben Fowkes. New York: Vintage Books, 1977.

Mbembe, A. "Necropolitics." Translated by Libby Meintjes *Public Culture* 15.1 (2003): 11-40.

McAfee, K. "Corn Culture and Dangerous DNA: Real and Imagined Consequences of Maize Transgene Flow in Oaxaca." *Journal of Latin American Geography* 2.1 (2003): 18–42.

McCaffery, Daniel F., Daniel Koretz, J. R. Lockwood, and Laura S Hamilton. "Evaluating Value-Added Models for Teacher Accountability." Santa Monica: RAND Corporation, 2005.

McLaren, P. *Capitalists and Conquerors: A Critical Pedagogy against Empire.* Lanham, MD: Rowman & Littlefield, 2005.

Mead, J. F., and P. C. Green III. "Chartering Equity: Using Charter School Legislation and Policy to Advance Equal Educational Opportunties." *National Education Policy Center.* 2012. http://nepc.colorado.edu/publication/chartering-equity

Merck, Sharp & Dohme Inc. "Fostering Science Education: Advancing the Dialogue toward a Healthier Future." 2009. http://www.msd-belize.com/corporate-responsibility/research-medicines-vaccines/science-education/home.html.

Meyer, R. H. "Value-Added Indicators of School Performance: A Primer." *Economics of Education Review* 16.3 (1997): 283–301.

Mills, C. W. *Blackness Visible: Essays on Philosophy and Race.* Ithaca, NY: Cornell University Press, 1998.

———. *The Racial Contract.* Ithaca, NY: Cornell University Press, 1997.

Mincer. J. *Schooling, Experience, and Earnings*. New York, NY: Columbia University Press, 1974.

Molnar, A. *School Commercialism: From Democratic Ideal to Market Commodity*. New York, NY: Routledge, 2005.

Montgomery, D. R. *King of Fish: The Thousand-Year Run of Salmon*. Cambridge, MA: Westview Press, 2003.

Moynihan, R., I. Heath, and D. Henry. "Selling Sickness: The Pharmaceutical Industry and Disease Mongering." *British Journal of Medicine* 324 (2002): 886–891.

Nadesan, M. "Governing Autism: Neoliberalism, Risk, and Technologies of the Self." *Governmentality Studies in Education*, edited by M. Peters, A. C. Besley, M. Olssen, S. Mauer, and S. Weber, 379–396. Rotterdam, The Netherlands: Sense Publishers, 2009.

National Academies of Sciences (NAS). *Rising above the Gathering Storm: Energizing and Employing America for a Brighter Economic Future*. Washington D.C.: National Academies Press, 2007.

———. *A Framework for Science Education: Preliminary Public Draft*. 2010. http://www7.nationalacademies.org/bose/Standards_Framework_Homepage.html.

National Center on Performance Incentives (NCPI). "About NCPI." 2011. http://www.performanceincentives.org/about-ncpi/index.aspx.

National Economic Council. "A Strategy for American Innovation: Driving toward Sustainable Growth and Quality Jobs." 2009. http://www.whitehouse.gov/administration/eop/nec/StrategyforAmericanInnovation/.

———. "A Strategy for American Innovation: Securing Our Economic Growth and Prosperity." 2011. http://www.whitehouse.gov/innovation/strategy.

National Institute for Mental Health (NIMH). "NIMH Mission." 2012. http://www.nimh.nih.gov/about/index.shtml.

National Research Council Board on Testing and Assessment. Letter Report to the U.S. Department of Education on Race to the Top Fund. Washington DC: National Academies Press, 2009.

Noble, D. *American by Design: Science, Technology, and the Rise of Corporate Capitalism*. New York: Oxford University Press, 1977.

Nowotny, H., P. Scott, and M. Gibbons. *Re-Thinking Science: Knowledge and the Public in the Age of Uncertainty*. Malden, MA: Polity Press, 2004.

O'Conner, C. "Denver's Aquaponics Project Aims to Turn 'Food Desert' into an Oasis of Health." *Denver Post*. 2010. Retrieved August 6, 2011, from http://www.denverpost.com/ci_13586527.

Olson, J. *The Abolition of White Democracy*. Minneapolis, MN: University of Minnesota Press, 2004.

Olssen, M. *Michel Foucault: Materialism and Education*. Westport, CT: Bergin & Harvey, 1999.

Olssen, M., and M. A. Peters. "Neoliberalism, Higher Education, and the Knowledge Economy: From the Free Market to Knowledge Capitalism." *Journal of Education Policy* 20.3 (2005): 313–345.

Orfield, G., G. Siegel-Hawley, and J. Kucsera. "Divided We Fail: Segregation and Inequality in the Southland's Schools." Los Angeles, CA: The Civil Rights Project Proyecto Derechos Civiles, 2011. http://civilrightsproject.ucla.edu/research/metro-and-regional-inequalities/lasanti-project-los-angeles-san-diego-tijuana/divided-we-fail-segregated-and-unequal-schools-in-the-southfield.

Patel, R. *Stuffed and Starved: The Hidden Battle for the World Food System*. Brooklyn, NY: Melville House Publishing, 2008.

———. *The Value of Nothing: How to Reshape Market Society and Redefine Democracy.* New York: Picador, 2009.

Pateman, C., and C. Mills. *Contract and Domination.* Malden, MA: Polity Press, 2007.

Pedretti, E. and J. Nazir. "Currents in STSE Education: Mapping a Complex Field, 40 Years On." *Science Education* 95 (2011): 601-626.

Peña, D. G., ed. *Chicano Culture, Ecology, Politics: Subversive Kin.* Tucson, AZ: University of Arizona Press, 1998.

Peters, M. "Education, Enterprise Culture and the Entrepreneurial Self: A Foucauldian Perspective." *Journal of Educational Enquiry* 2.1 (2001): 58–71.

———. "Governmentality, Education, and the End of Neoliberalism?" In *Governmentality Studies in Education*, edited by M. Peters, A. C. Besley, M. Olssen, S. Maurer, and S. Weber, xxvii–xlvii. Rotterdam, Amsterdam: Sense Publishers, 2009.

———. "The New Prudentialism in Education: Actuarial Rationality and the Entrepreneurial Self." *Educational Theory* 55.2 (2005): 123–137.

Peters, M. A., and T. Besley. *Subjectivity and Truth: Foucault, Education, and the Culture of Self.* New York: Peter Lang Publishers, 2007.

Peters, M., A. C. Besley, M. Olssen, S. Maurer, and S. Weber, eds. *Governmentality Studies in Education.* Rotterdam, Amsterdam: Sense Publishers, 2009.

Pharmaceutical Research and Manufacturers of America (PhRMA). *R&D Investment by U.S. Biopharmaceutical Companies Remains Strong Despite Ongoing Economic Challenges.* 2010. Retrieved May 13, 2012, from http://www.phrma.org/news/news/rd_investment_us_biopharmaceutical_companies_remains_strong_despite_ongoing_economic_chall.

Phillips, C. B. "Medicine Goes to School: Teachers as Sickness Brokers for ADHD." *PLoS MED* 3.4 (2006): e182.

Pierce, C. "Designing Intelligent Knowledge: Epistemological Faith and the Democratization of Science." *Educational Theory* 57.2 (2007): 123-140.

Pietila, A. *Not in My Neighborhood: How Bigotry Shaped a Great American City.* Lanham, MD: Ivan R. Dee Publishers, 2011.

Plomin, R., and F. M. Spinath. "Intelligence: Genetics, Genes, and Genomics." *Journal of Personality and Social Psychology* 86.1 (2004): 112–129.

Popkewitz, T. S., and M. Brennan, eds. *Foucault's Challenge: Discourse, Knowledge, and Power in Education.* New York: Teacher's College Press, 1997.

Prasad, A. "Capitalizing Disease: Biopolitics of Drug Trials in India." *Theory, Culture & Society* 26.1 (2009): 1–29.

Rabinow, P. *Essays on the Anthropology of Reason.* Princeton, NJ: Princeton University Press, 1996.

———. "Studies in the Anthropology of Reason." *Anthropology Today* 8.5 (1992): 7–10.

Rabinow, P., and N. Rose. "Biopower Today." *Biosocieties* 1.2 (2006): 195–217.

Robelen, E. "STEM Education Gets Boost from Race to Top Winners." *Education Week.* 2010. http://blogs.edweek.org/edweek/curriculum/2010/08/stem_education_to_get_big_boos.html.

Roediger, D. *The Wages of Whiteness: Race and the Making of the American Working Class.* New York: Verso Press, 1991.

Rose, N. *Governing the Soul: The Shaping of the Private Self.* London: Free Association Books, 1999.

———. *The Politics of Life Itself: Biomedicine, Power, and Subjectivity in the Twenty-first century.* Princeton, NJ: Princeton University Press, 2007.

Roth, W. M., and A. C. Barton. *Rethinking Scientific Literacy*. New York: Routledge, 2004.

Rudolph, J. L. *Scientists in the Classroom: The Cold War Reconstruction of American Science Education*. New York: Palgrave Macmillan, 2002.

Saltman, K. *Capitalizing on Disaster: Taking and Breaking Public Schools*. Boulder, CO: Paradigm Publishers, 2007a.

———. Ed. *Schooling and the Politics of Disaster*. New York: Routledge, 2007b.

———. *Education as Enforcement: The Militarization and Corporatization of Schools*. New York: Routledge, 2010.

———. *The Edison School: Corporate Schooling and the Assault on Public Education*. New York: Routledge, 2006.

Sanders, W. L. Statement of William L. Sanders Before the Committee on Education and the Workforce Hearing on "No Child Left Behind: Can Growth Models Ensure Improved Education for All Students." 2007. Retrieved December 13, 2011, from http://archives.republicans.edlabor.house.gov/archive/hearings/109th/fc/nclb072706/sanders.htm.

Sanders, W. L., and Sandra P. Horn. "The Tennessee Value-Added Assessment System (TVAAS): Mixed-Model Methodology in Educational Assessment." *Journal of Personnel Evaluation in Education* 8 (1994): 299–311.

Schools for Chiapas. 2012. http://www.schoolsforchiapas.org/.

Schultz, T. *Investment in People: The Economics of Population Quality*. Berkeley, CA: University of California Press, 1981.

Shaw, P., J. Lerch, D. Greenstein, W. Sharp, L. Clasen, A. Evans, J. Giedd, F. X. Castellanos, and J. Rapoport. "Longitudal Mapping of Cortical Thickness and Clinical Outcomes in Children and Adolescents with Attention-Deficit/Hyperactivity Disorder." *Archives of General Psychiatry* 63 (2006): 540–549.

Shear, M., and N. Anderson. "Obama Uses Funding to Pressure Education Establishment for Change." *The Washington Post*. 2009. Retrieved October 2, 2011, from http://www.dc.gov/DCPS/In+the+Classroom/Ensuring+Teacher+Success/IMPACT+%28Performance+Assessment%29/Value-Added.

Shiva, V. *Biopiracy: The Plunder of Nature and Knowledge*. Boston, MA: South End Press, 1997.

———. *Earth Democracy: Justice, Sustainability, and Peace*. Boston, MA: South End Press, 2005.

Shukin, N. *Animal Capital: Rendering Life in Biopolitical Times*. Minneapolis, MN: University of Minnesota Press, 2009.

Sidorkin, A. M. "Human Capital and the Labor of Learning: A Case Study of Mistaken Identity." *Educational Theory* 57.2 (2007): 159–170.

Silver, L. M. *Remaking Eden: How Genetic Engineering and Cloning Will Transform the American Family*. New York: Avon Books, 1997.

Silver, M., and M. Pugh. (dir.) *The Leech and the Earthworm*. Indigenous Peoples Council on Biocolonialism Films DVD, 2003.

Simons, M. "Learning as Investment: Notes on Governmentality and Biopolitics." *Educational Philosophy and Theory* 38.4 (2006): 523–540.

Singer, E. "A Neurological Basis for ADHD: Scientists Have Identified a Genetically Determined Pattern of Brain Development Linked to ADHD." *Technology Review*. http://www.technologyreview.com/news/408381/a-neurological-basis-for-adhd/

Singh, I. "Bad Boys, Good Mothers, and the 'Miracle' of Ritalin." *Science in Context* 15, no. 4 (2002): 577–603.

Snider, V. E., T. Busch, and L. Arrowood. "Teacher Knowledge of Stimulant Medication and ADHD." *Remedial and Special Education* 24.1 (2003): 46–56.

Soling, C. *The War on Kids.* Bronx, NY: Spectacle Films, 2009.

Stanley, W., and N. Brickhouse. "Teaching Sciences: The Multicultural Question Revisited." *Science Education* 85.1 (2001): 35–49.

Stanworth, M. "Reproductive Technologies and the Deconstruction of Motherhood." In *Reproductive Technologies: Gender, Motherhood, and Medicine*, edited by M. Stanworth. Cambridge: Polity Press, 1987.

Steinberg, Ted. *Down to Earth: Nature's Role in American History.* Oxford: Oxford University Press, 2008.

Stern Stewart & Company. "Company History." 2011. http://www.sternstewart.com/?content=history&p=2000s.

Strauss, V. "Hawaii Teachers Reject Contract in 'blow to Race to the Top." *The Washington Post.* 2012. Retrieved May 21, 2012, from http://www.washingtonpost.com/blogs/answer-sheet/post/hawaii-teachers-reject-contract-in-blow-to-race-to-the-top/2012/01/20/gIQA2KHCGQ_blog.html.

Sunder Rajan, K. *Biocapital: The Constitution of Postgenomic Life.* Durham, NC: Duke University Press, 2006.

Tauli-Corpuz, V. "Biotechnology and Indigenous Peoples." In *Redesigning Life?: The Worldwide Challenge to Genetic Engineering*, edited by B. Tokar, 252–270. New York: Zed Books, 2001.

Taylor, J. *Making Salmon: An Environmental History of the Northwest Fisheries Crisis.* Seattle, WA: University of Washington Press, 1999.

The MTA Cooperative Group. "A 14-Month Randomized Clinical Trial of Treatment Strategies for Attention-Deficit/Hyperactivity Disorder." *Archives of General Psychiatry* 56.12 (1999): 1073–1086.

Thompson, C. *Making Parents: The Ontological Choreography of Reproductive Technologies.* Cambridge, MA: MIT Press, 2007.

Thrush, C. *Native Seattle: Histories from the Crossing Over Place.* Seattle, Washington: University of Washington Press, 2007.

Tippins, D. J., M. P. Mueller, M. van Eijck, and J. D. Adams, eds. *Cultural Studies and Environmentalism: The Confluence of Ecojustice, Place-based (Science) Education, and Indigenous Knowledge Systems.* New York: Springer, 2010.

Turque, B. "Rhee Dismisses 241 D.C. Teachers; Union Vows to Contest Firings." *The Washington Post.* 2010. http://www.washingtonpost.com/wp-dyn/content/article/2010/07/23/AR2010072303093.html.

Turque, B., and N. Anderson. "Delaware, Tennessee Win Education Awards in First Race to the Top Competition." *The Washington Post.* 2010. http://www.washingtonpost.com/wp-dyn/content/article/2010/03/29/AR2010032901276.html.

United Nations Food and Agricultural Organization. *The State of World Fisheries and Aquaculture.* Rome, Italy: Communication Division, 2008.

United Nation's International Narcotics Control Board. "Psychotropic Substances: Statistics for 2009." 2010. http://www.incb.org/incb/psychotropics_reports.html.

United States Department of Education. "Nine states and District of Columbia Win Second Round Race to the Top grants." 2010. http://www.ed.gov/news/press-releases/nine-states-and-district-columbia-win-second-round-race-top-grants.

———. "Race to the Top Fund." 2012. http://www2.ed.gov/programs/racetothetop/awards.html.

Valdéz, J. G. *Pinochet's Economists: The Chicago School of Economics in Chile*. New York: Cambridge University Press, 2008.

Waldby, C. *The Visible Human Project: Informatic Bodies and Posthuman Medicine*. New York: Routledge, 2000.

Watkins, W. H. *The White Architects of Black Education: Ideology and Power in America, 1865–1954*. New York: Teachers College Press, 2001.

Welch, C. "Meet Salmon Farming's Worst Enemy: A Determined Biologist." *Seattle Times*. 2012. http://seattletimes.nwsource.com/html/localnews/2018296338_virus-lady27m.html.

White, R. *The Organic Machine: The Remaking of Columbia River*. New York: Straus and Giroux, 1995.

Wilkinson, C. *Messages from Frank's Landing: A Story of Salmon, Treaties, and the Indian Way*. Seattle, WA: University of Washington Press, 2000.

Willinsky, J. *Learning to Divide the World: Education at Empire's End*. Minneapolis, MN: University of Minnesota Press, 2008.

Willis, P. *Learning to Labor: How Working Class Kids Get Working Class Jobs*. New York: Columbia University Press, 1981.

Winerip, M. "As Best Schools Compete for Best Performers, Students May be Left Behind." *New York Times*. 24 July 2011. http://www.nytimes.com/2011/07/25/nyregion/at-best-schools-competing-for-best-performers-students-may-be-left-behind.html?pagewanted=all.

Wong, J. "Growing Dendrites: Brain-Based Learning, Governmentality and Ways of Being a Person." In *Governmentality Studies in Education*, edited by M. Peters, A. C. Besley, M. Olssen, S. Maurer, and S. Weber. Rotterdam, Amsterdam: Sense Publishers, 2009.

Zeidler, D. L., T. D., Sadler, M. L. Simmons, and E. V. Howes, eds. "Beyond STS: A Research-Based Framework for Socioscientific Issues Education." *Science Education* 89.3 (2005): 357–377.

Zizek, S. "Have Michael Hardt and Antonio Negri Rewritten the *Communist Manifesto* for the Twenty First Century?" *Rethinking Marxism: A Journal of Economics, Culture, & Society* 13.3 (2001): 190–198.

Index

Printed in the United States of America